CAMBRIDGE LIBRARY COLLECTION

Books of enduring scholarly value

Cambridge

The city of Cambridge received its royal charter in 1201, having already been home to Britons, Romans and Anglo-Saxons for many centuries. Cambridge University was founded soon afterwards and celebrated its octocentenary in 2009. This series explores the history and influence of Cambridge as a centre of science, learning, and discovery, its contributions to national and global politics and culture, and its inevitable controversies and scandals.

Handbook to the Natural History of Cambridgeshire

This early twentieth-century guide to the geography and geology, fauna and flora of Cambridgeshire was written during a period when natural history played a particularly prominent role in British cultural life. The heart of the book is a comprehensive survey of the diversity of animal life in the region, focussing particularly on the insect orders. It also includes chapters on vertebrate palaeontology and archaeology. Two maps show locations of discovery of ancient skulls, as well as important ancient roads that cross the county. There are additional botanical and geological maps. The book provides a valuable baseline for present-day studies of biodiversity or the effects of climate change, and will also appeal to local enthusiasts with an interest in environmental history.

Cambridge University Press has long been a pioneer in the reissuing of out-of-print titles from its own backlist, producing digital reprints of books that are still sought after by scholars and students but could not be reprinted economically using traditional technology. The Cambridge Library Collection extends this activity to a wider range of books which are still of importance to researchers and professionals, either for the source material they contain, or as landmarks in the history of their academic discipline.

Drawing from the world-renowned collections in the Cambridge University Library, and guided by the advice of experts in each subject area, Cambridge University Press is using state-of-the-art scanning machines in its own Printing House to capture the content of each book selected for inclusion. The files are processed to give a consistently clear, crisp image, and the books finished to the high quality standard for which the Press is recognised around the world. The latest print-on-demand technology ensures that the books will remain available indefinitely, and that orders for single or multiple copies can quickly be supplied.

The Cambridge Library Collection will bring back to life books of enduring scholarly value (including out-of-copyright works originally issued by other publishers) across a wide range of disciplines in the humanities and social sciences and in science and technology.

Handbook to the Natural History of Cambridgeshire

JOHN EDWARD MARR
ARTHUR EVERETT SHIPLEY

CAMBRIDGE
UNIVERSITY PRESS

CAMBRIDGE UNIVERSITY PRESS

Cambridge, New York, Melbourne, Madrid, Cape Town, Singapore,
São Paolo, Delhi, Dubai, Tokyo

Published in the United States of America by Cambridge University Press, New York

www.cambridge.org
Information on this title: www.cambridge.org/9781108007665

© in this compilation Cambridge University Press 2009

This edition first published 1904
This digitally printed version 2009

ISBN 978-1-108-00766-5 Paperback

THE

NATURAL HISTORY

OF

CAMBRIDGESHIRE

London: C. J. CLAY AND SONS,
CAMBRIDGE UNIVERSITY PRESS WAREHOUSE,
AVE MARIA LANE.

Leipzig: F. A. BROCKHAUS.
New York: THE MACMILLAN COMPANY.
Bombay and Calcutta: MACMILLAN AND CO., Ltd.

HANDBOOK

TO THE

NATURAL HISTORY

OF

CAMBRIDGESHIRE

EDITED BY

J. E. MARR, Sc.D., F.R.S.,

AND

A. E. SHIPLEY, M.A., F.R.S.

CAMBRIDGE:

at the University Press

1904

Cambridge:

PRINTED BY J. & C. F. CLAY,

AT THE UNIVERSITY PRESS.

PREFACE.

IT is the object of this Guide to afford help to those students of Natural History who desire to make observations in the Cambridgeshire district.

It is our pleasant duty to acknowledge the assistance we have received from many willing helpers in preparing the Guide.

In the first place we wish to tender thanks to the various authorities on the different subjects who have without remuneration written the articles included in the book.

To the Editors and Publishers of the Victoria History of the Counties of England we owe our thanks for the article on the Vertebrate Palæontology of the County which was written by Mr R. Lydekker for the County History of Cambridge, and is here allowed to appear in advance. We are also indebted to them for the use of other materials acknowledged in the body of the work.

The Geological Map of the County was kindly made by Mr H. H. Thomas, M.A., of H.M. Geological Survey. Our thanks are due to him and also to the Controller of H.M. Stationery Office, who has permitted us the use of the topographical details inserted on the map of H.M. Ordnance

Surveys (4 miles to the inch), and of the Geological lines which are taken from the maps of H.M. Geological Survey. The map has been prepared at the Office of the Ordnance Surveys, Southampton.

The other maps were prepared for the Guide by Mr Edwin Wilson, of Cambridge.

To Mr R. H. Rastall, B.A., of Christ's College, we offer our sincere thanks for the compilation of the Index.

Lastly, we desire to record our obligations to the Syndics of the University Press for their liberality in presenting copies of the work to members of the British Association, and to others connected with the Press for the courteous manner in which they have on all occasions assisted us.

J. E. M.
A. E. S.

CAMBRIDGE,
June, 1904.

CONTENTS.

MAPS.

ADDENDA AND CORRIGENDA.

p. 114, line 2. For "101" read "102."

p. 115, line 18. For "many places" read "some places."

p. 117 *et seqq.* For "peregra" read "pereger" (*v.* B. B. Woodward, J. Conch., Oct. 1903, p. 362).

p. 120, line 27. For "*Vitrea nitida*" read "*Zonitoides nitidus.*"

p. 120, line 32. After "*Testacella haliotidea*" insert "found by Mr H. Watson."

p. 120, line 33 should read "It was introduced from Kew about twenty years ago by Mr R. I. Lynch."

p. 121, line 3. For "a common form" read "a form found here and there."

p. 121, line 8. Add "the first two in the tropical fern house and the last in the water lily pond. These introductions were discovered by Mrs Hughes."

p. 123, line 19. Add "Both the records are by Mrs Hughes."

p. 123, line 25. After "occurs" add "both species being found first by Mrs Hughes."

p. 127, line 18. For "*cristata*" read "*costata.*"

p. 128. Add to Pelecypoda, Cardiidae, *Cardium edule*, Linn.

p. 131, line 4. After "colony" insert "by Mrs Hughes."

p. 131, line 21. Add "including a tumid variety found by Mr Watson."

p. 133, line 8. For "Reid" read "Reed."

p. 133, line 25. For "they contain" read "Mr Jukes-Browne records (*v.* Mem. Geol. Survey 'The Geology of the Neighbourhood of Cambridge' pp. 88, 107)."

p. 133, line 30. For "Somersham" read "Bluntisham."

p. 138, line 19. Add "This discovery was made recently by Mrs Hughes, who obtained in the same bed shells of *Cardium* and *Macoma* as well as of several non-marine species."

p. 138, line 31. After "Trinity College" insert "Mr F. F. Laidlaw, M.A., Trinity College, and Mr Hugh Watson."

p. 216, line 15. For "commencement" read "end."

THE

PHYSIOGRAPHY OF CAMBRIDGESHIRE.

By J. E. MARR, Sc.D., F.R.S., and
W. G. FEARNSIDES, M.A.

THE county of Cambridge is about fifty miles in length
from north to south, and its greatest breadth from east to
west is about thirty miles; it occupies an area of eight
hundred and sixty square miles. The greater part of the
county is drained by the river Ouse and its tributaries, of
which the Cam is the principal, while the northern part is
traversed by a portion of the Nen and its tributaries.

Along part of its course the old Ouse river ran in a
direction somewhat north by east from St Ives, past Ely,
dividing the county into two portions, of which the northern
is almost entirely composed of fenland, while the southern is
largely occupied by more elevated ground.

In accordance with its geological structure the county is
divisible into three important areas, and two minor ones.
The former consist of (i) the chalk tract which lies to the
east of the Cam between the southern part of the county and
Waterbeach, (ii) the curiously dissected plateau which occupies
the south-western part of the county between the valleys of
the Cam and Ouse, and (iii) the Fenland of the northern
part of the county. The minor tracts consist of the alluvial
belts which border the Cam and its tributaries and the ridges
of old river gravels; and a small plateau topped by gravels
which occupies the country around Fordham, Chippenham,
and Newmarket.

The full significance of the various features will be grasped when the reader has studied the section of this guide which is concerned with an account of the geology of the county, but it is necessary to refer at this point to the general characters of the geological formations underlying the area.

Of the three important areas into which the county is structurally divisible, the first-mentioned only can be said to possess a form which is the direct outcome of the influence of the stratified rocks which underlie it. In the second area the underlying strata apparently exercise little effect upon the character of the surface features. Chalk, Gault clay, Greensand and Jurassic clays alike rise to a plateau-like surface, which for a distance of fifteen miles does not depart from the two hundred and twenty foot contour to an extent of more than ten or a dozen feet. This plateau is much dissected by agents of denudation, and the shape of its contours is distinctly remarkable. The third area, the Fenland, owes its position to the existence of a great thickness of slightly inclined Jurassic clays beneath it, but the actual form and flatness of the area has been determined, firstly by denudation of these clays, probably as the combined effect of subaërial and marine action, and secondly by deposit of silt and by plant-growth.

The more ancient rocks which exist in the county belong to the Jurassic and Cretaceous systems. As the strata dip gently in a direction a little south of east, the older (Jurassic) rocks occupy the northern and western portions of the county, while the southern and eastern parts are occupied by the newer (Cretaceous) rocks. The Jurassic rocks consist chiefly of clays, and are therefore marked, on the whole, by low ground.

The nature of the Cretaceous rocks varies. The lowest group consists of sands ; these are succeeded by clays which crop out to the east of the sands, and east of the clays lie the various members of the chalk.

The superficial accumulations which cover the more ancient rocks are very variable. The principal, in addition to ordinary surface soil, are Boulder-clay, gravels, sands and loams, and peat.

The drainage of the rivers is to some extent determined by the direction of strike of the strata, but in the Fens many changes have taken place in recent times, and the meandering courses of the river have again and again been altered. The course of the Ouse through the county is, as a whole, that of a strike stream, though the portion between Bluntisham and Ely, known as the Old River, flows in the general direction of dip of the strata. The Rhee, from its source near Ashwell to its junction with the Granta, and beyond this to Bottisham, is also a strike stream. One important tributary from the west, the Bourn, flows in the direction of the dip to the junction with the Cam near Trumpington. Most of the tributaries, however, flow in a direction opposite to that of the dip of the strata, and drain the chalk tract. The most noteworthy are the Granta, which flows from Saffron Walden, the Lin, from Bartlow, the Lark, from Bury St Edmunds, and the Little Ouse from Thetford. All of these have low passes (wind-gaps) at their head, and may occupy valleys which were once drained by streams flowing in a direction contrary to that of the present streams.

Though the pre-glacial courses of the principal streams were much as now, many minor modifications were produced during and after glacial times, but into this matter we cannot enter here.

We may now proceed to give a brief account of the characters of the principal and minor areas which we have previously defined.

(i) *The Chalk Tract.* The main tract of Chalk occupies that part of the county which lies east of the Rhee, and of the Cam after the Rhee has joined it. It is bordered on the west by wide sheets of gravel, and its higher hills are capped by Boulder-clay. Several of the hills in the south-east of the county rise above the four hundred foot contour-line, and much of this district is relatively elevated. The district consists

1—2

chiefly of scarps facing a little north of west, and of dip
slopes facing south of east. The various divisions of the
Chalk form scarps of different degrees of importance, the
principal being those of the Lower Chalk, which is well seen
at Cherryhinton, and of the Upper Chalk seen near Balsham.
The shape of the hills is that of typical chalk downs, namely
rolling undulations, and where the Boulder-clay has been
removed the vegetation is also characteristic, though the flora
of low-growing plants has been greatly modified as the result
of cultivation. Until the beginning of the nineteenth century
the ground was largely occupied by short turf, but most of
it is now under the plough and the character of the original
turf is only preserved here and there, as on the Fleam Dyke and
the Devil's Dyke. Lines of beech-trees mark the boundaries
of the chalk. A good deal of the Boulder-clay land has been
planted with trees, and the chalk country as a whole is quite
well wooded and is distinctly picturesque. The district, as
above noted, is drained by a series of streams flowing in a
direction opposite to that of the dip of the strata. Many
dry valleys show that the superficial drainage was once more
extensive : now, the bulk of the rain which falls is absorbed,
and much of it is given out by a series of springs which mark
the outcrop of the impervious Chalk-marl at the base.

(ii)   *The Western Plateau.* This is, geologically speaking,
the most complicated of the Cambridgeshire districts. It
is mostly covered with a thin capping of Boulder-clay, but
in places the Boulder-clay is of great thickness. Occasionally
the underlying rock is exposed on the low ground.

The district is bounded on the south-east by the outcrop
of Gault along which the Rhee flows. It includes the Chalk
outliers of Madingley and of the Haslingfield-Wimpole region,
with the large tract of Gault to the west of them. The Lower
Greensand from Gamlingay to Lolworth and the Upper Jurassic
clays to the west of that also underlie this plateau. Most of
the rocks are therefore clays, and the character of the country
is much the same whether Boulder-clay occurs at the surface

or is absent. An exception to this is seen when the ground
is occupied by uncovered Chalk or Greensand, each of which
determines the existence of its peculiar flora. Neither,
however, exerts much influence upon the form of the plateau,
which exists as high land because it forms the watershed
between the Ouse and Cam, though the exact conditions
under which the plateau structure was brought about have
yet to be determined.

The district is flanked on the north-east by a great finger
of river-gravel which extends out into the Fen basin along
the line of the Huntingdon Road between Girton and Fen
Stanton.

(iii) *The Fenland.* Generally speaking, the county to the
north of a line joining Over and Newmarket is occupied by
fen. 'Islands' project here and there, as at Ely, Upware,
and March, but these form only a very small proportion of
the whole tract. The Fenland occupies a wide-spread hollow
which was excavated in pre-glacial times, for Boulder-clay is
often found beneath the fenland accumulations. Its width
is due to the gentle arching over of the belt of Jurassic Clays,
the erosion of which has determined the existence of the
fenland tract. The floor is probably not a very even surface,
but silt deposited by the sea, mud brought down by the rivers,
and vegetation growing upon the surface have built up the
area to a very uniform plain not more than a foot or two
above mean-tide level. The Fens, therefore, in pre-Roman
times had become " a swampy plain interspersed with drier
places, and with deep morasses," and in the parts near the sea
this was liable to be flooded by high tides.

The tract is largely occupied by peat in the south, and
by silt in the north: the latter occupies a strip of country
about twelve miles in width just south of the Wash, and
extends southwards as a great triangular wedge with its
apex near Littleport.

The conditions of the Fenland have greatly altered during
historic times as the result of human interference. The

Romans built a series of great sea-banks to keep out the tides, and in so doing interfered with the river outlets, which in course of time became choked with silt, and overflowing their banks rendered the area more of a morass than before.

In medieval times many artificial drainage works were undertaken to improve matters, but most of them were performed in so ill-advised a way that only small areas were favourably affected. In the early part of the seventeenth century a new system was inaugurated, and the natural waterways have since been opened and embanked, with the consequence that pumping has been reduced to a minimum, and the district has been rendered ideal for farming operations.

With the disappearance of the boggy fen many of the characteristic plants and animals have also disappeared or become extremely rare. A small area at Wicken Fen is still preserved in its original state and shows the former character of extensive tracts of Fenland.

As the Fenland has undergone so much change in recent times it is interesting to find a description of the Isle of Ely and adjoining region as it was in the twelfth century in the *Liber Eliensis*, of which the MS. is preserved in the library of Trinity College. The particular passage quoted below is from a translation furnished by a reviewer in the *Zoologist* for 1879 (Third Series, Vol. III., p. 71). "In our isle men are not troubling themselves about the leaguer, but think they may safely be defended by their tiros; the ploughman has not taken his hand from the plough, nor has the hunter cast aside his arrow, nor does the fowler desist from beguiling birds. And yet something more. If you wish to hear what I have known and have seen, I will reveal all to you. The isle is within itself plentifully endowed, it is supplied with various kinds of herbage, and for its richer soil surpasses the rest of England. Most delightful for its charming fields and pastures, it is also remarkable for its beasts of chase, and is in no ordinary way fertile in flocks and herds. Its woods and vineyards are not worthy of equal praise, but it is beset by great meres and fens as though by a strong wall. In this isle there is an abundance of domestic cattle and a multitude of wild animals; Stags, Roes, Goats, and Hares are found in its groves and by these fens. Moreover there is a fair plenty of Otters, Weasels, and Polecats, which in a hard winter are caught by traps, snares, or by any other device But what am I to say of the kind of fishes, and of fowls, both those that fly and those that swim? In the eddy at the sluices of these meres are netted innumerable Eels, large Water-wolves—even Pickerels, Perches, Roaches, Burbots, and Lampreys, which we call Water-snakes. It is indeed said by many men that sometimes *Isicii*[1], together with the royal fish, the Sturgeon, are taken. As to fowls, let us, if it be not troublesome to you, recount those which abide there and thereabout, as we have done with the rest. There are numberless Geese, *Fiscedulae*, Coots, Didappers, Water-crows, Herons, and Ducks, of which the number is very great. At midwinter or when the birds moult their quills I have seen them caught by the hundred, and even by three hundreds more or less, sometimes they are taken in nets and snares as well as by bird-lime[2]."

(iv) *Minor Tracts.* The river banks are usually marked by a strip of alluvium of varying width, which gives rise to flat ground often flooded in wet weather. This alluvium varies in width. Below Cambridge it passes into the flat of the fenland near Waterbeach.

The ancient alluvia, deposited when the rivers ran at a high level, form terraces more or less parallel to the present river-courses, but often departing from strict parallelism, as will be described more fully in the section devoted to geology. They occur in all the valleys of the present rivers ; also in certain now streamless valleys on the Chalk, and as long

---

[1] The reviewer suggests that this word means salmon, and the suggestion is confirmed by another writer in the same volume of the *Zoologist*, p. 222.

[2] Lib. ii. cap. 105 (ed. D. J. Stewart, 1848).

irregular ridges protruding from the edges of the Fen. They consist chiefly of fragments of flint.

In the neighbourhood of Fordham, Chippenham, and Newmarket a sheet of gravel of uncertain origin caps a plateau between fifty and one hundred feet above the adjoining fenland. Its origin is yet unexplained. It is chiefly of interest on account of its flora. This is specially marked in the old gravel-pit of Chippenham, as noticed by Prof. C. C. Babington. The plants appear to have settled there, owing to the similarity between this gravel and the sandy deposits of East Norfolk and Suffolk, though those deposits have an origin different from that of the Chippenham gravels.

The climate of Cambridgeshire is summed up in the Geological Survey Memoir on "The Geology of the Neighbourhood of Cambridge" as follows:—

" The physical conformation of the ground is (with the exception of the chalk slope) such as to induce dampness of soil and atmosphere ; a characteristic which, as regards Cambridgeshire, has become proverbial. It is owing, not to the rainfall which in amount is small, about 23 inches a year, but to the low-lying area being surrounded by higher ground on all sides but one, by the preponderance of ground sloping to the north, and by the prevalence of clay soils."

THE GEOLOGY OF CAMBRIDGESHIRE.

By W. G. FEARNSIDES, M.A., F.G.S.

THE geology of Cambridgeshire, so far as the underlying strata are concerned, is not complicated; the superficial deposits, however, exhibit considerable variety, and their origin is still in many cases a subject for discussion.

The stratified rocks of Cambridgeshire form part of the great mass of secondary strata which extends in an unbroken line from the Yorkshire coast to that of Dorsetshire. In the tract of which Cambridgeshire forms a part, their general strike is nearly north-east and south-west, and as the beds dip gently towards the south-east (at angles which approach horizontality) the older strata lie to the north and west of the county and the newer ones to the south and east. To the north-west of a line drawn from near Littleport to Gamlingay the rocks mainly belong to the Jurassic system; to the south-east of that line to the Cretaceous System.

Complications are introduced by folding, unconformabilities, and the existence of outliers. An anticlinal fold brings up a considerable mass of Jurassic rocks in the neighbourhood of the hamlet of Upware. An important unconformity occurs at the base of the Cretaceous rocks, causing the lower Cretaceous rocks to rest upon different members of the Jurassic System, while a smaller unconformity occurs at the base of the Chalk.

Several outliers of Lower Cretaceous rocks rest upon the Jurassic rocks to the west of the main line of outcrop of the Lower Cretaceous group, and outliers of Chalk repose on lower members of the Cretaceous System to the west of the main development of the Chalk. The Jurassic rocks, with the exception of a local development near Upware, consist of slightly consolidated clays or muds, while the Cretaceous rocks present a greater variety of sediments. At the base is an iron-stained sand, sometimes compacted into sandstone. This is succeeded by a thick deposit of clay, and at the top is the Chalk. No Tertiary sediments occur in the county, so our knowledge of events during that period depends entirely on the record contained in the rocks of other areas.

The only accumulations newer than the chalk are of post-Tertiary date. They consist of Boulder-clay, accumulated during the 'Great Ice Age,' and of subsequently formed gravels of fluviatile and marine origin, and of the alluvia of the river-valleys and the peat and silt of the Fenland.

For works on the geology of the county the reader may be referred to : (i) the Rev. T. G. Bonney's *Cambridgeshire Geology*, published in 1875 ; (ii) F. R. C. Reed's *Handbook to the Geology of Cambridgeshire*, which appeared in 1897 ; (iii) *The Geology of the Neighbourhood of Cambridge*, being the explanation of Quarter-sheet 51 S.W. and part of 51 N.W., by W. H. Penning and A. J. Jukes-Browne, published in 1881 ; and (iv) *The Geology of the Fenland*, by S. B. J. Skertchly, which appeared in 1877. The two latter are official publications of H.M. Geological Survey. The list of geological survey maps and sections which illustrate the geology of the county will be found in Mr Reed's bibliography, which forms the appendix to his work on Cambridgeshire geology. This bibliography also gives a very full list of the various books and pamphlets which have appeared at different times, containing references to the geology of the county.

We may now proceed to describe the various deposits of the county in order, commencing with the oldest strata.

JURASSIC.

Oxfordian.

The oldest rocks of Cambridgeshire belong to that division of the Upper Oolites which is known as the Oxford Clay. This clay covers a very considerable area in the north and west of the county, but much of the tract which it occupies is drift-covered, and artificial exposures are rare. The total thickness of the Oxford Clay hereabouts has been variously estimated at from 400 to 600 ft. A well was sunk at Bluntisham to 300 ft., but did not reach the base of the deposit.

The lower beds may be seen about Peterborough, in the adjoining county. The *Belemnites Owenii* beds are there extensively worked for bricks, and a number of reptilian bones and Ganoid fish, as well as Belemnites and other Mollusca, have been obtained from them.

At Whittlesea, in Cambridgeshire, there are large brick works in similar beds, which possibly appertain to a higher horizon.

In the more accessible pits of Huntingdonshire and Bedfordshire much higher beds are worked, and the resulting bricks are white or yellow instead of red.

At Eynesbury, near St Neots, in Huntingdonshire, the beds with the *ornatus* group of ammonites are well exposed : they are grey or grey-blue-clays, which on weathering become creamy. They are underlain by much darker clays, which weather brown, but these are apparently not so profitable to work and are badly exposed. Thin bands of marly or sandy limestone occur at intervals through the clay, and it is from these or from associated beds that the best preserved ammonites are obtained. One bed recently exposed was a mass of *Ammonites athleta*, the individuals ranging up to a foot and a half across.

Sandy (Bedfordshire) is another well-known locality for fossils. Here, a little beyond the station, higher beds are worked,

and yield an abundance of ammonites of the *cordatus* group,
especially *A. cordatus* and *A. Mariae,* together with Belemnites of various species and a great many lamellibranchs. The
stiffer and darker clays of the bottom of the pit contain much
selenite. Two well-marked limestone bands are seen, but
should be avoided by the fossil hunter, being most unprofitable.
The best fossils are preserved in pyrites, and cavities within
them often contain beautiful crystals of barytes. In the
cutting between the pit and the station the two limestone
bands may be seen cut off by the Lower Greensand unconformity.

At St Ives (Huntingdonshire), in Saint's brick pit, one mile
from the town, at the junction of the Chatteris and Ely roads,
beds quite similar to those at Sandy are well exposed, and here
also are very fossiliferous. The famous pit on the river side
north-east of the town is quite overgrown, and little but the
rubble fallen from the St Ives rock which topped the exposure
can now be seen. Two bands of sandy limestone were at one
time seen interbedded with the highest beds of Oxford Clay,
and that which appears near the top of Saint's pit is probably
the lower of these bands.

At Holywell were formerly large brickyards which were
specially rich in fossils, and a well recently put down near the
ferry showed similar fossiliferous beds. In this well the
true character of the highest Oxford Clay was seen : it is a
very blue tenacious clay, which on weathering becomes light
grey-blue and ultimately passes into an almost creamy mud.
It is moderately rich in selenite and occasionally very
pyritous.

The commonest fossils of this district are *Ammonites
cordatus, A. Mariae, Belemnites hastatus, Waldheimia im-
pressa, Rhynchonella nucula,* and numerous lamellibranchs, as
Arca, Nucula and *Gryphaea.*

At Godmanchester is another pit which shows beds somewhat lower than those at St Ives ; fossils are there scarce.

Corallian.

Lying comformably on the top of the Oxford Clay over most of the district is a somewhat remarkable bed which has been correlated by Mr T. Roberts and others with the Lower Calcareous Grit of Yorkshire. This bed varies somewhat from place to place along its outcrop which is much obscured by drift. It has been named Elsworth Rock and St Ives Rock. It is a hard calcareous or ferruginous mudstone, but differs from the so-called limestone bands of the Oxford and Ampthill Clays in that it contains iron-shot oolitic grains which are apparently composed of chalybite. When fresh, as in well sections, the bulk of the rock is blue, but the oolitic grains are always brown. When weathered the whole rock becomes rusty brown with occasional creamy patches where oolite grains are not too abundant. The rock is almost always exceedingly fossiliferous.

The best and, at present, the only workable exposure, is in the village street at Elsworth and in the stream which runs along that street. Until recently good material could be obtained from one of the farmyards at the south end of the village, but now the erection of a building has obscured that exposure. Mr Wedd in mapping the country on the six inch scale has been able to trace the outcrop of the Elsworth Rock over a very considerable distance, from Meadow Farm, south of Croxton, by Yelling and Papworth Everard to Elsworth and on to Red Hill Farm. He also noted a probable exposure between Fen Drayton and Swavesey, and north of the Ouse found a fairly continuous outcrop from Holywell by Lindsell's Farm and the old St Ives brickyards as far west as Houghton Hall.

About Elsworth the rock may be divided into a lower limestone, four to seven feet thick; a middle black but rusty clay, five feet; and an upper limestone, one and a half feet. It is the upper limestone which is most constantly oolitic, the lower is often concretionary and may be hard and creamy.

The middle clay is often replaced by sandy beds and occasionally disappears altogether.

The commonest fossils are the ammonites of the *Plicatilis*
type, together with almost all the ammonites of the highest
Oxford Clay. *Gryphaea dilatata* is ever present, and the
same may almost be said of *Exogyra nana. Pholadomya,
Pecten, Avicula* and a few gasteropods, may generally be
obtained. *Collyrites bicordata* is the most characteristic
fossil, but there is no species which we can definitely record
as peculiar to the horizon.

Away from the line of outcrop, rocks agreeing with the
Elsworth rock are known in various well sections. Those at
Bourne and at Chettering Farm, seven miles due east of
Holywell and three from Upware, are most interesting. In
each, the three-fold division is recognizable and the oolitic
character is well marked. At Upware the rock reappears at the
surface near the Inn and has been traced for about a mile to
the east. A well just north of the Inn exhibited it overlain by
some three feet of coral limestone rubble, and Mr Wedd, by
digging at the bottom of a small pit in the adjoining field,
found more of it overlain by some seven feet of undisturbed coral
limestone which, from its lithological characters and included
fossils, he correlated with the well-known Coral Rag of the
south pit at Upware. More recently beds agreeing with the
upper beds have been recognized in a trial hole made through
the Coralline Oolite of the South pit, and it seems probable
that the Elsworth rock underlies all the Corallian deposits
of the Upware inlier.

Above the Elsworth rock over most of the district lies the
Ampthill Clay, a clay much like the Oxford Clay but differing
from it in several points of detail. This contains a rich but
ill-preserved fauna which has enabled Mr Roberts and others to
correlate it with the Corallian of other districts. The Ampthill
Clay is darker in colour than the Oxford Clay and contains
a good deal more carbonaceous material, and though pyritous
in places, rarely has its fossils preserved in pyrites. It seems

to have been deposited more slowly than the Oxford Clay, for all its shelly fossils are encrusted with growths of *Serpula* and it contains occasional phosphate nodules.

In the south and west of the district a good deal of the Ampthill Clay was removed by denudation before the deposition of the Lower Greensand and hence the sections there are incomplete. The greatest thickness is attained in the district about Over where it may be about one hundred and fifty feet thick. Towards Upware it thins rapidly and at Chettering is but twenty-six feet in thickness.

The best exposures are at Gamlingay, where Ampthill Clay is worked in three pits. Another pit is worked at Everton, west of Great Gransden, and the most westerly pit at Haddenham is probably in its upper layers.

The Belle Vue Brickyard at Gamlingay shows an interesting section. Flooring the pit is a hard calcareous bed whose lower layers contain enormous quantities of *Serpula*. This is supposed by many to be the local representative of the Elsworth rock. Some nine or ten feet above it is another hard bed, and above this some seven or eight feet brings us to the base of the Lower Greensand which unconformably overlies the Ampthill Clay. Following along the length of the section we are able to make out a definite transgression from east to west. The junction of clay and sand is absolutely sharp, but the last two or three feet of the clay are quite different in character from the rest, and are rejected by the brick-makers who call the material *Bung*. This 'bung' is evidently much crushed and slickensided and is occasionally rucked up into little anticlines which encroach upon the domain of the sand above.

The commonest and most characteristic of the fossils are the flattened oysters *Ostrea discoidea*. *Belemnites abbreviatus* and biplicate ammonites may generally be collected, and there are very many crushed Pectens and other lamellibranchs.

Towards Upware, as we have said, the Ampthill Clay thins, and gives place to the coral island deposits shown in the

Upware sections. These are of two types, Coralline Oolite
and Coral Rag. Of these the Oolite is in general supposed to
underlie the Rag, but in the now closed pit near the Inn
Mr Wedd found rock of the Rag type resting directly on
Elsworth Rock, and though each has its own distinctive fauna,
it is conceivable that here the order is not invariable, the
faunistic assemblage being merely a question of habitat.

The two exposures termed the North and the South pits
have long been known. The North pit is now almost entirely
overgrown, but the South pit has been so extended that
practically all that was ever seen in the North pit can now
be observed in its walls and floor, and a trip to Upware is still
one of the most instructive and enjoyable excursions that one
can make from Cambridge.

The South pit is about a quarter of a mile north of the
Inn and is about two hundred yards long. The beds dip at
about five degrees to the north and hence the lowest beds
exposed occur in the south corner of the pit. In 1901 a trial
hole and boring was put down near this corner. The first
four feet was dug, and pisolitic rock similar in all respects to
that of the quarry floor was passed through. At four feet
a hard bed, very like the Coral Rag of the other end of the
quarry, was met with, and when the first layer of this was
broken through, so much water sprang up that digging had to
be abandoned. A long bar was therefore obtained and a
boring made. This broke through several layers of coral
rock, intermediate marly beds and iron-stained rubble; and,
at a depth of nine feet, came to a very hard rock containing
iron-shot oolitic grains. This rock is believed to be the
Elsworth rock.

From the pisolite of the dug portion of the pit, *Gryphaea
dilatata*, many *Cidaris* spines, and a *Collyrites* were collected.
These were associated with the ordinary fossils of the Coralline
Oolite.

The Oolite is exposed over about three-quarters of the
length of the pit and passes gradually up into the Rag.

Typically the Oolite is a creamy and not very compact limestone, consisting of flattened or oval grains about half the size
of a pea. Hosts of tiny gasteropods and lamellibranchs are
represented by casts among the grains, and only large calcite
shells survive in their entirety. *Gervillea, Opis Mytilus* and
Pecten are the commonest fossils, but in certain beds, especially near the top of the series, sea-urchins such as *Holectypus
depressus, Echinobrissus scutatus* and *Pygaster umbrella*
abound.

Above the Oolite comes the Rag, which consists chiefly
of recrystallized tabular coral growths. It is platy, full of
holes and very hard, and is in fact a typical coral reef. The
chief reef builders were corals of the genera *Isastraea, Mont-
livaltia, Stylina* and *Thamnastraea*, and these are associated
with very numerous lamellibranchs such as are common even
now on coral reefs, e.g. *Opis, Arca* and *Lithodomus*. Gasteropods also are abundant, and urchin fragments are by no
means rare. *Cidaris florigemma* is the most plentiful.
Ammonites are rare.

Occasionally when Gault is being dug at the south-east
corner near the river entrance, another mass of Rag is to be
seen. It immediately underlies the Lower Greensand, and
must, I think, be identical with the rag and rubbly beds
found in the boring, as its position is such that unless some
unseen fault intervenes it cannot overlie the main mass of
Oolite.

The total thickness of limestone exposed in the pit must
be about fifty feet, of which about twenty belongs to the
Coral Rag, and the rest to the Oolite and underlying beds.

The North pit is about one and a half miles north of this
main pit : no fresh rock can now be seen there, but urchins
can still often be picked out among the rubble near the
water's edge. The beds are very similar to the middle beds
of the Oolite of the South pit and the faunas are identical.
The beds of the North pit still dip to the north at a low
angle, and hence must undulate considerably in the interval

between the pits. Such undulation can also be abundantly
proved by a consideration of the well sections at some of the
farms.

Limestone beds like those about Upware extend all along
the ridge nearly to Barway—a total distance of about three
miles. The width of the ridge is about one mile, and a well
boring near Wicken has proved the absence of limestone
there.

Kimeridgian.

Overlying the Ampthill Clay in all places where that
formation is present in its entirety is the Kimeridge Clay.
South of Knapwell the whole of it had been removed before
the deposition of the Lower Greensand, but further north
the diminution of the unconformity allows the coming in
of higher and higher beds, and a consequent widening of the
outcrop.

The Kimeridge Clay appears to lie conformably on the
Ampthill Clay, and a line of phosphatic nodules is taken as
the horizon at which to separate the two. At Upware, at the
north-east end of the inlier, Prof. W. Keeping found that the
Kimeridge Clay lay unconformably against a bank of Corallian limestone, but no exposure of such a section has since
been seen.

The phosphate nodule bed has been observed in the now
overgrown pit at Knapwell, and in a pond and ditch west of
Oakington, also near Willingham, Haddenham Fen, and in the
brickyard half a mile west of Haddenham station, which is
still worked down to the bed in question. It is also known
in the Chettering boring. The clay immediately above the
nodules at Haddenham is very dark in colour and has an
extraordinary abundance of *Ostrea deltoidea,* both large and
small.

The Kimeridge Clay, like the underlying Ampthill and
Oxford Clays, is variable in character ; it is in general darker
and more finely laminated than either. In its upper portions

it becomes so bituminous as to be unfit for brick-making. Thin bands of limestone are fairly common, but are inconstant. They often pass into lines of septarian nodules and are generally unfossiliferous.

The best exposure is that shown in the Roswell Pit, one mile north-west of Ely, where all but the very lowest Ostrea beds are at times exposed. Some seven feet below the ordinary water-level of the Ouse is a thin band of fissile sandy shale with crowds of *Astarte supracorallina*. Above that is about ten feet of blue clay becoming sandy and papery towards the top. A few calcareous nodules are met with about this horizon, and just above them Aptychi of ammonites abound. Twelve feet of variable grey sandy shales follow, and above them darker shales crowded with *Exogyra virgula*. This horizon is well marked by a band of large septaria, and above it come seven feet of the very bituminous shales with *Discina latissima*. These are the highest Jurassic beds known in the district, and they are overlain by somewhat disturbed Lower Greensand. Of other exposures those at Haddenham are interesting as showing the lowest beds of the clay, which are ill exposed at Ely. The pit to the west of the station shows the basement phosphate bed, that south of the station shows somewhat higher beds. Various exposures of the intermediate layers also occur in three pits along the Ely-Littleport road and in several brickyards at Littleport, but these add but little to the knowledge obtained from the section at the Roswell Pit.

The Kimeridge Clay is generally quite rich in fossils. Flattened ammonites and lamellibranchs are to be obtained from nearly every layer, but solid ones are somewhat rare. Few of the fossils are pyritized, but some are preserved as phosphate nodules, and saurian bones in that condition are not rare. The common ammonites are *biplex, mutabilis, alternans* and *calisto*.

CRETACEOUS.

Lower Greensand.

Towards the end of Jurassic time a notable uplift affected the whole area of East Central England, and the old Jurassic sea area was broken up.

South Cambridgeshire seems to have been near the summit of the rising arch, but it is not quite clear at what period the area was first affected. Certain it is, however, that uplift was not persistent through the whole time interval indicated by the unconformity. Kimeridge and Lower Portland Clays show little change in lithological character all along their outcrop, and Mr Lamplugh and others in working out the relations of the Speeton, Tealby and Snettisham Clays and the Spilsby and Sandringham Sandstones have shown that each successive zone of these Neocomian strata overlaps its predecessor on to the flank of the uplifted ridge. A similar state of affairs is probable among the more estuarine beds of the south, but much work remains to be done before such overlap can actually be proved there.

Uplift was followed by a gradual sinking of the now denuded ridge, and towards the end of Lower Greensand times the sea was able to break across it and again link the water areas of the north and south.

Under such conditions the Lower Greensand deposits of Cambridgeshire were laid down, and it is not therefore surprising that they are somewhat anomalous in character.

The Lower Greensand of Cambridge consists of a variable thickness of clean bright yellow or yellow-brown sands, much false bedded on both large and small scales. When brought up from deep borings it is found to contain numbers of green-coated sand grains and occasional glauconite concretions which seem to be casts of foraminiferal chambers. The sand is almost entirely a quartz sand; its grains are generally sub-angular, but the larger ones have their edges rounded and their surfaces well polished.

The cement is oxide or sulphide of iron, but except near the surface the sand is generally quite friable. Near the surface percolating water has largely re-distributed the cementing material, and depositing it along certain beds and joints has brought about the formation of the hard Carstones which give the rock its local name. The irregular occurrence of the same process has given rise to the celebrated 'boxstones' of these beds.

One of the strangest features of our Lower Greensand is the presence of beds of rolled phosphatic nodules or coprolites among it. These have long attracted the attention of geologists, and, being at one time of considerable economic importance, were also available for careful study. The beds of phosphate come in at various horizons and are rarely continuous over wide areas. They form conglomerates, and their nodules, many of which are rolled casts of recognizable fossils, bear in themselves evidence of being derived from the various Upper Jurassic Clays. A few are of later date, but Mr Teall declares that in certain places more than 50 per cent. were once portions of *Ammonites biplex*. Whatever the source, the phosphate content of the nodules seems to be practically invariable. Along with the nodules are many pebbles of such rocks as vein quartz, chert, lydian stone ; and boulders of gneiss, granite, mica schist, and trilobite-bearing Carboniferous limestone have been observed.

In most places the Lower Greensand is unfossiliferous, but wherever phosphate beds have been worked, a few, and in some places a great many, indigenous fossils have turned up. Fragments of coniferous wood are generally abundant. The present exposures of Lower Greensand in Cambridge are poor in the extreme. The outcrop over most of its course is completely buried by superficial deposits such as Boulder-clay, gravel and peat, and north of Cambridge the thickness developed is too small to exert any distinctive influence on the character of soil or scenery.

At Sandy, in Bedfordshire, some three miles beyond the

county boundary, the Greensand is about 120 feet thick, and a fine cliff section, some 60 feet high, may be seen on the north side of the railway just west of the station. Here it rests directly on Oxford Clay, and it is interesting to note that within an inch of the junction the sand has no trace of clayey material. The Greensand beds exposed are fine yellow sands, much false bedded, but they contain no important phosphate band. A little higher on the hill, however, an important phosphate bed came to the surface and has been worked from a large portion of the area known as Sandy Heath.

The railway from Sandy to Potton skirts along the Lower Greensand scarp which there exhibits its characteristic scenery and flora.

At Potton, about three-quarters of a mile south of the station, a pit is still worked for phosphates but is 'preserved' with great care and is well-nigh unapproachable. Under favourable circumstances fine examples of derived phosphates pierced by indigenous boring animals may be collected; the commonest are in the form of casts of *Ammonites biplex*, but lamelli-branchs and reptilian bones are by no means rare. Erratic pebbles are not very common.

The phosphate bed is overlain by about seven or eight feet of very brown sand, with occasional carstones, and the exposure is topped by a wash of Boulder-clay.

Within the county boundary the Greensand may be seen resting on Ampthill Clay in the three pits at Gamlingay, but no very interesting features can there be made out.

At Great Gransden and also at Little Gransden exposures showing good Carstone are also visible, and that at Great Gransden is particularly interesting on account of the enormous scale of its false bedding. The exposure is capped by Boulder-clay which rests on an absolutely even surface which obliquely truncates the false-bedding planes. No trace of Greensand pebbles is here seen in the Boulder-clay.

North-east of Great Gransden the Greensand is completely buried by drift, and except for a small stream-exposure at

Bourn, is never seen at the surface until one gets to near
the Cambridge-Huntingdon road between Dry Drayton and
Lolworth. There a strip of sandy land occurs and is continued by Oakington and Cottenham to the Fens. A similar
strip rests upon Kimeridge Clay north of the Ouse and runs by
Haddenham and Wilburton to Stretham, where it was once
worked for the phosphates which occur scattered through its
mass. Ely, too, is built on an outlier of Greensand which,
though much drift covered, is exposed at the north corner of
the Roswell pit.

Around the Upware 'island' a thin representative of the
Lower Greensand occurs. Its greatest thickness is not much
more than ten feet, but the locality was unsurpassed for
fossils and phosphates. The Greensand of Upware rests
upon Coral Rag, Coralline Oolite, and Kimeridge Clay. Its
lowest layers consist of angular lumps of the limestones set in
a clayey matrix. A pebbly sand with abundant phosphate
nodules follows, and in places has a calcareous cement, and
the various indigenous brachiopods swarm in it. Above this
comes a layer of yellow sand and then another phosphate bed,
not so rich in fossils but still containing numerous lamellibranchs, gasteropods, and sponges. Another foot or two of
sand completes the series, which is topped by a layer of clay
and another nodule bed containing Gault fossils.

Unfortunately phosphate working at Upware and Wicken
has come to an end, and the only available exposure of
these interesting beds is that furnished by a shallow ditch
on the east side of the green road which runs from the South
to the North pit at Upware. The best exposure is about
300 yards north of the South pit. There the rock may be
examined and occasional oysters, *Terebratulidae* and *Rhyncho-
nellidae* can be collected.

The section at the river entrance to the main South pit
when dry shows the relations of the beds rather well.

The study of the indigenous faunas found at these various
localities led Mr Teall in 1875 to refer the Lower Greensand

of Cambridgeshire to a late stage in the Lower Greensand period. Mr W. Keeping in 1883 fully confirmed this result, and still more recently (1903), Messrs Lamplugh and Walker, examining a limestone band near the top of the Lower Greensand of Leighton Buzzard, Bedfordshire, have found a fauna which according to them has strong affinities with that of the Upper Greensand of some parts of England.

Gault.

Overlying the Lower Greensand is the Gault.

The mutual relations of the two are not always very clear, but in Cambridgeshire there seems to be complete conformity with overlap by the higher beds in all directions.

A rapid subsidence in late Neocomian times had brought Lincolnshire and Norfolk within the area of pelagic conditions, and, had not the supply of sediment from the south been comparatively large, Cambridgeshire, too, would have shared in those conditions; and as it is, a very notable change in the thickness of the Gault is observable within the limits of the county. Thus well-borings at Ashwell have proved nearly 200 feet of Gault clay, while at Cambridge there is but 150 feet, at Soham 90, and at West Dereham, just over the Norfolk border, only 60 feet. Still further north, at Roydon, there is only about 19 feet of clay with a two-foot band of yellow and red marls in the middle of it, and at Dersingham the whole is represented by less than ten feet of the red marls, and between that place and Heacham these pass into the well-known Hunstanton Red Rock, which though only three feet thick represents more than the two hundred feet of clay found in a district less than 50 miles distant.

This change of thickness, great as it is, doubtless does not represent the whole change really involved. Between Soham in Cambridgeshire and Barton in Bedfordshire, the Gault is overlain by Cambridge Greensand, a bed which contains many phosphate nodules believed to have been locally derived from the Gault by subaqueous denudation, and it is generally stated

that the Upper, and in places, the higher part of the Lower
Gault has been used up in the process and is therefore absent
in the district.

The Gault of Cambridge is a stiff tenacious clay, dull
leaden grey in colour and with a tinge of brown in it. On
drying it becomes much paler but still retains the same dead
hue. It varies little from bed to bed and contains many
indigenous phosphate nodules sporadically distributed. It is
generally rather unfossiliferous, but certain of its beds are by
no means so, and occasional *Plicatulae* and *Belemnites* can be
picked up from most of its exposures.

The lower part of the Cambridge Gault is never well
exposed. Most of its outcrop is covered by drift and has
never yet been exploited for brick-making. At Upware it is
occasionally dug on the river side of the South pit for use
in bank making, and in these circumstances the lowest layers
may be easily seen. The Gault here rests directly on Lower
Greensand and appears to overlap it on to the old coral reef.
Its lowest layers are glauconitic and somewhat sandy in
texture. *Belemnites attenuatus* and *B. minimus* abound, and
the nodules of a phosphate bed, about a foot from the base,
consist largely of *Ammonites interruptus*; *Nucula* and *Inoceramus sulcatus* are also plentiful.

Similar fossiliferous beds are found all over the county
at the base of the Gault, and well-sinkers accordingly hail
with joy the appearance of fossils in their diggings.

The higher part of the Cambridge Gault is well exposed
and can be examined at leisure in the fine sections provided
by the brickyards of Barnwell, and also at Barrington. Less
interesting exposures are open about a mile from Cambridge
along the Barton Road, on the Ely Road, at Clayhithe, and at
Soham. The upper surface is often bared in coprolite workings
and in many places where the Chalk Marl is being worked for
cement.

Derived and somewhat worn fossils from the Upper Gault
are extremely abundant in the Cambridge Greensand, and the

highest beds of the local Gault are remarkably devoid of determinable fossils. Hence, and on theoretical grounds, it has been generally supposed that the Upper Gault proper was in Cambridge entirely denuded soon after its deposition.

A recent examination of the large pit, east of the New-market Road at Barnwell, has proved the existence of a richly fossiliferous horizon, some forty feet below the Cambridge Greensand, and from this any quantity of *Ammonites varicosus* may be collected. The fossil belt exposed is about seven feet thick and has not yet been pierced ; it contains occasional brown contemporaneous phosphate nodules throughout. One of the beds is in places so rich in phosphate nodules and shell fragments that it has to be rejected by the brickmakers as 'blowing' the bricks, and it is exceedingly hard and tough. The phosphate nodules occur in various states, some being hard and compact while others are merely lumps of partly phosphatized mud. All are unworn and are apparently contemporaneous with the rest of the deposit. Grains of glauconite are abundant, and these, with the shell fragments, give the rocks the characteristic feel of certain beds of the Upper Greensand of the south. The commonest fossils of the band are ammonites, and examples of *A. varicosus* may be observed up to a foot in diameter, and *A. splendens* is also fairly abundant. A large concentrically striated *Inoceramus* often three or four inches across is very plentiful and provides most of the calcareous material for the hard bed. Other interesting forms as *Terebratula biplicata, Terebratulina triangularis, Pecten orbicularis* and *Nucula bivirgata* are common. The whole fauna presents an aspect very similar to that of the Upper Gault nodules of the Cambridge Greensand. From this new evidence, therefore, we conclude that 'Upper Gault' is not entirely absent in the neighbourhood of Cambridge, but that at least forty of the one hundred and fifty feet proved in well-sections at Barnwell must be ascribed to the Upper Gault.

The section at Barrington adds confirmatory evidence in the

same direction. The whole section will be described later; in the meantime it is interesting to note that there too the Gault is fossiliferous and, though ammonites are rare, the common Upper Gault lamellibranch *Avicula gryphaeoides* is abundant.

Cambridge Greensand and Chalk Marl.

About the end of the Gault epoch a further swelling of the arch which had determined the Neocomian unconformity of Central England brought the sea floor of the Cambridge area again within the region of effective denudation, and a slight unconformity between Gault and overlying Chalk Marl resulted.

As is usual in such cases, a pebble bed rests upon the eroded surface of the Gault, and as this is of economic importance and is of a type somewhat unusual among similar deposits, it has received a special name, the Cambridge Greensand.

The great feature of the bed is the unusual abundance of slightly worn black phosphate nodules which have been washed out of the Upper Gault. These form about ninety-five per cent. of the pebbles in the basal bed and the richness of the fauna provided by them is remarkable.

The remaining five per cent. of the pebbles are mostly contemporaneous brown phosphate nodules, and these contain a fauna proper to the Chalk Marl. Occasional lumps of foreign rock, granite, gneiss, porphyry, basalt, purple grit, quartzite, etc., rounded or angular, and up to a foot in diameter, have also been recorded. The pebble bed varies from a few inches to about a foot and a half in thickness; it rests on a pockety surface of Gault, the phosphate nodules or Coprolites being most abundant in the hollows.

The matrix is mostly a chalky marl but is much richer in clayey material than the overlying beds. Grains of glauconite supposed to be derived from denuded Gault are abundant both among the pebbles and also for a foot or two above them, and it is these that have earned for the bed its name Greensand.

Passing up from the Greensand we find a continuous diminution of the proportion of clay and a corresponding increase in the amount of chalk, and this continues right up to the top of the Chalk Marl.

The natural mixture of chalk and clay is very favourable to the making of cement, and the gradation of composition enables manufacturers who judiciously mix their different beds to obtain cement of constant composition; they are also able to vary that composition, and so make the different grades of cement.

Unfortunately these manufacturers, when they obtain cement chemically similar to more famous cements which are known by names which declare the stratigraphical horizon of their origin, do not hesitate to apply these stratigraphical names to their Chalk Marl cement, and unless this be stopped much confusion is likely to ensue.

The thickness of the Chalk Marl in South Cambridge is about sixty feet and the whole is worked for cement. Northward it thins gradually, and at Cambridge only the lower thirty feet is marly enough for cement making. At Hunstanton, the whole is only sixteen feet thick and has become a very pure white limestone—the Sponge Bed.

Exposures of Chalk Marl are very numerous in Cambridge, and were formerly much more so. Generally speaking, the whole outcrop of the Cambridge Greensand has been worked over and its coprolites extracted to a depth of about twenty feet, but all these workings are now closed and it is only where cement pits happen to be worked to the base of the marl that we are able to examine its peculiarities.

At the pit of the Royston Cement Company, about a mile south of Barrington, the Chalk Marl is worked for cement and the underlying Gault for bricks, and as these are worked separately a considerable shelf of the greensand and phosphate bed is left exposed.

The exposure in the more easterly pit is most interesting, but the western one shows the higher as well as the lower

Chalk Marl. The eastern pit is topped by some six to twelve feet of river gravel containing mammalian bones. This will be described later. It rests on a firm but irregular surface of the lower Chalk Marl, here quite yellow in colour. Fossils are not abundant in the marl but *Terebratula semiglobosa, Ostrea vesicularis, Lima,* and *Inoceramus* may generally be picked up. The Greensand bed below is typically rich in fossils but is now no longer washed. Its commonest fossils are sharks' teeth, *Avicula gryphaeoides, Plicatula sigillina, Terebratula biplicata, Terebratulina triangularis, Belemnites minimus, Ammonites rostratus* and *A. splendens.* Most of these occur both as pebbles and as contemporaneous fossils, but the ammonites are all derived. The Gault below is exposed for about thirty feet; it is quite normal in its character and, as above mentioned, contains numerous lamellibranchs.

Another interesting exposure is that of the Mill Road Cement Works, Cambridge. There the Greensand is taken out as exposed and, when sufficient accumulates, is washed for its coprolites. At certain times good material can be collected as of old. Gasteropods are the speciality of the locality and the crabs *Eucorystes* and *Palaeocorystes* are not rare. The Greensand is overlain by some twenty feet of marl. At Hauxton Mill is another very similar pit which, in times past, has yielded many fine fossils. The Chalk Marl is here more fossiliferous than usual and yields *Kingena lima* and occasional casts of ammonites.

Of other exposures we may note those of the cement works on both sides of the Great Northern Railway, near Shepreth station and to the east of it. Also the various pits at Meldreth, Bassingbourne, Orwell, and Madingley, but none of these exhibit complete sections or show the relationship of the beds to their surroundings.

Chalk Marl disturbed by the extraction of the coprolite bed at its base may also be seen in all the above-mentioned brickyards of the Gault.

Lower or Grey Chalk.

Resting conformably on the Chalk Marl is the Grey Chalk or Zone of *Holaster subglobosus*. The basal beds of this series are of somewhat unusual character and are variously known as Totternhoe Stone, Burwell Rock or Clunch. These beds differ markedly from the rest of the Chalk of the district in that they contain an appreciable proportion of detrital quartz. They are very compact and harsh to the touch, but this is due to the presence of a cement among the shell fragments rather than to the actual quartz grains. The rock is well and regularly jointed, and lying as it does between porous Chalk above and impervious Chalk Marl below generally determines the line of springs which follows the base of the Chalk escarpment. The rock is rich in phosphate, and green lumps of phosphatized chalk are common, especially at the base.

The only exposure of the lowest beds is to be found in the large pit nearest the station at Burwell, where the rock is still worked as a building stone. The higher beds are also worked in this pit and in another some four hundred yards to the north of it and are burnt for lime. Fossils are most abundant in the lowest beds, but the limeburners collect from the higher ones only. The most characteristic fossils are the grinders and other teeth of sharks, the various Pectens—*Beaveri, orbicularis,* and *fissicosta,* several species of *Inoceramus, Terebratula semiglobosa, Rhynchonella Mantelliana,* and *Holaster subglobosus.* The rock is twenty feet thick at Burwell but varies much.

Another tolerable exposure is still open just south of Haslingfield and there the peculiar *Ostrea frons* and various ammonites may be collected.

At one time the Totternhoe stone was worked all along its outcrop, but there is now no further demand for soft building-stone suitable only for interior decoration, so the working has ceased and the pits are much overgrown.

The beds immediately above the Burwell Rock have many features in common with that rock but are less compact and contain far less detrital matter. At Cherryhinton the first twenty feet is considered unfit for lime-burning, but near Fulbourn Asylum and at Swaffham it is quarried for that purpose. Pits opened in these beds are generally worked for clunch or wattle making. We may mention those of Newton, Eversden, Haslingfield, and Shelford. The upper fifty feet of the *subglobosus* zone is much whiter than any of the beds below it, and is a hard homogeneous chalk in which the stratification is not at all obvious. Being very pure and yet quite free from flints it is locally much used in the making of lime for building, and the sections at Cherryhinton are the best in the district.

The great pit of Cherryhinton is situated at the north end of a spur of the Gog Magog Hills and, having been worked well into the hill, it exhibits in continuous series all the members of the Lower Chalk.

The older north eastern end of the pit is no longer worked, but the upper beds of Burwell Rock and the overlying clunch beds can still be seen in the deep hole on the right of the Cherryhinton entrance. Fossils may be collected there, the commonest being *Rhynchonella Martini, Terebratula biplicata,* and the lamellibranchs *Inoceramus, Pecten,* and *Lima.*

Lime-burning from the rest of the quarry is proceeding vigorously, but the beds are poor in fossils and it is rare to meet with anything determinable. The men working in the lower part of the pit, however, sometimes come across specimens of a few species and save all they can get. The most usual are *Terebratula semiglobosa,* and beautiful specimens of the zone fossil *Holaster subglobosus.* Sharks' teeth and *Pecten orbicularis* are not rare, and the big *Pecten Beaveri* occurs. The whole recorded fauna does not exceed a score of species.

The higher part of the pit seems to contain no fossils, but shows very clearly the peculiar curved jointing characteristic

of the beds. Having no stratification along which the successive layers may slip when slightly folded, the rock had to
develope horizontal joints before the normal formation of the
dip and strike series could occur. These horizontal joints,
therefore, resemble tiny thrust planes, and as such are curved
rather than straight, so when stretching joints commenced
to form, they too were somewhat irregular, and now the whole
effect is strongly suggestive of false-bedding. Contraction
during the drying and later irregular solution by percolating
water may also have had their effect.

The exposure is capped by a few feet of the so-called
Belemnite marls and Melbourn Rock, which here are not
particularly interesting and contain no fossils. They have,
however, a yellowish colour, and so serve to indicate the throw
of a couple of small faults recognizable below by their slickensided surfaces and fault breccia.

Another large pit in beds similar to those of Cherryhinton
is opened on the Caius Golf Links, by the side of the Gog
Magog Road. Fossils are not abundant, but *Holaster* and
occasional *Pecten* may be obtained from the workmen.

Above the *Holaster subglobosus* zone is the zone of
Belemnitella (Actinocamax) plena, here represented by some
seven feet of alternating sandy marls and hard yellow chalk.
The marly beds contain lumps of apparently older white chalk,
and the surface of the chalk below is said to be eroded.
Fossils are rare.

The line separating these marls from the overlying hard
Melbourn Rock is taken as marking the division between the
Lower and Middle Chalk.

Middle or White Chalk.

The lowest zone of the Middle Chalk is the zone of *Rhynchonella Cuvieri*, and in Cambridge this zone is believed to
include the Melbourn Rock.

The Melbourn Rock is a band of very hard greenish-yellow chalk full of even harder nodules of white chalk which is often recrystallized. It is very persistent, and though never more than ten feet thick is sufficient to determine a feature along the Chalk escarpment.

The rest of the zone consists of ordinary white chalk with very occasional bands of flint nodules. It has numerous marly partings and is well stratified throughout. *Rhynchonella Cuvieri* is generally abundant and swarms in certain layers. The thickness assigned is sixty to seventy feet.

The best exposure of Melbourn Rock is in the lime pit at Melbourn. There it has its usual character and rests on the Belemnite marls. A similar exposure, much overgrown, exists at Steeple Hill, Shelford. Fossils are always rare.

The beds immediately above the Melbourn Rock are well seen in the old pit and antiquarian excavations on the south-east side of the road above the great Cherryhinton pit, in which *Rhynchonella* and *Inoceramus labiatus* swarm. Other localities worth noting are the road cutting about a hundred yards north-east of the Babraham cross roads, and a small pit on the east side of the road from Stapleford to the Gog Magog Hills. The first of these is rich in *Rhynchonellae*, but the second shows only bedded chalk with *Ostrea vesicularis*.

The second zone of the Middle Chalk is known as the zone of *Terebratulina gracilis*, but the name is unfortunate, as the original of *T. gracilis* seems only to occur in the Norwich Chalk, while the proper name of the zone fossil is *T. lata*. The exposures near Cambridge are poor in the extreme, but being harder than its neighbours the zone determines a line of steeper slopes, and so is easily traceable along the escarpment. It is often termed the Wandlebury Chalk and is about one hundred feet thick. Two quarries are worked in it at Royston, and the old railway at Worsted Lodge provides a section. Fossils are badly preserved and rare. Flints are not abundant, but when present possess remarkable shapes and generally lie vertically rather than horizontally.

Upper Chalk or Chalk with Flints.

The next zone of the Chalk is that of *Holaster planus*, and this is now referred to as the lowest division of the Upper Chalk. It includes the well-known Chalk Rock, and also certain other beds of nodular white chalk both above and below that horizon.

The beds below the Chalk Rock are soft and white. Flints are very common at certain levels, but there are often wide intervals between the layers. Many of the flints contain sponges. Marl layers also occur and are especially numerous in the passage beds to the Chalk Rock.

The Chalk Rock consists of a variable number of hard semi-crystalline bands of creamy chalk which have layers of green phosphatized nodules and rubbly chalk between them. The upper surface of the rock is sharply defined, and may in places be a surface of contemporaneous erosion. This surface has long been taken as the base of the Upper Chalk, but Mr Jukes-Browne in comparing the rich fauna of the Chalk Rock with that of higher beds, finds that no separation of stages is here possible, so he includes the whole of the Planus zone in the Upper Chalk.

The beds above the Chalk Rock are generally soft and earthy, but harder bands occur and there are some marly layers. Beds of flints are not numerous, but tabular flints following joints are fairly common.

Exposures are more abundant than in the underlying beds, but all are at some distance from Cambridge.

The lowest beds may be well seen in a pit by the roadside just east of Great Chesterford, and there the zone fossil and many Micrasters and *Inoceramus* may be collected. Of other localities we may mention the little pit near Linton Workhouse, also Lark's Hall, a pit east of Six Mile Bottom, and lastly Dullingham.

The Chalk Rock is even more sought after, and old pits are common all along its outcrop. It is comparatively rich in fossils. The best collecting grounds are Reed, Barkway and Barley, south of the county boundary, but Westley Waterless, Underwood Hall, Wood Ditton, and especially Cheveley, in the east of the county, are also good. The fauna at all these places is large, and is especially notable for its aberrant ammonites and gasteropods. *Scaphites Geinitzi* and *Heteroceras* are characteristic. *Pleurotomaria, Turbo, Trochus* and *Natica, Rhynchonella plicatilis, Terebratula carnea* and several species of *Micraster* abound. Fossils are generally very difficult to extract, and weathered specimens are often the best.

The beds above the Chalk Rock yield far more specimens of *Micraster* than of *Holaster,* and accordingly were placed by the Geological Surveyors in the zone of *Micraster corbovis,* which has since been correlated with the zone of *Micraster cortestinudinarium.* Recently, however, specimens of the Micrasters have been sent to Dr Rowe and have been recognized by him as belonging to the '*Planus* zone,' so we now have to include a considerable thickness of this higher chalk in the zone of *Holaster planus.*

The area occupied by the surface outcrop of these beds within the county is quite small, for they are almost entirely covered by Boulder-clay. Fine exposures, however, do occur, and from many of them most beautiful specimens of the various fossils may be collected. The best localities are Balsham, a pit north-west of West Wratting, and the Middle of the World near Horseheath.

The most accessible is the Balsham lime pit. There most beautiful examples of *Spondylus spinosus* with all its spines are abundant, and uncrushed examples of *Micraster, Rhynchonella plicatilis* and *Terebratula carnea* occur. Occasional giant specimens of *Holaster planus* have been found.

The West Wratting pit is very rich in Micrasters but other fossils are rather rare.

The highest members of the chalk series of Cambridgeshire probably belong to the *Micraster cortestinudinarium* or

3—2

possibly to the *M. coranguinum* zone, but no exposures of
either can be definitely recognized within the county.

Eastward in Suffolk and in Norfolk the higher zones come
on in turn, and there is no reason for doubting that the whole
series was deposited over Cambridge. All has now gone, and
moreover, all had gone before the next succeeding deposit,
the Boulder-clay, was accumulated.

EOCENE—RECENT.

What occurred in the Cambridgeshire area in the interval
between Chalk and Boulder-clay we may never know, and at
present can hardly conjecture. Certain it is, however, that
a period of earth movement affected all England at the end
of Cretaceous time, and from being the bottom of a land-
locked gulf our area must have passed into the condition of
a coastal plain which bordered some vast continent stretching
far away to the north-west. In Eocene time a great river
from the west brought down the various deposits of the
London and Hampshire basins, and a slight sinking about
its estuary allowed the tides to bring Bagshot sands within
some fifteen miles of Cambridge.

The Oligocene and Miocene were periods of great disturbance, and the earth movements in giving the Mesozoic
rocks of England their present strike must have produced
profound changes in the pre-existing lines of drainage, and
may have given the dip streams their present courses.

Pliocene time brought with it a certain amount of subsidence, and a North Sea opening northward soon washed
over the Eastern Counties. A river from the south, an
ancestor of the present Rhine, brought down silt to be
deposited in that sea, and the Crags of the East Coast were
formed as shell banks beyond its delta, each series a little
further to the north than its predecessors.

Meanwhile the climate had been steadily becoming colder,
and by the end of the Pliocene period an almost Arctic flora
and fauna was living on the present shores of East Anglia.

Pleistocene followed Pliocene and the rigour of the climate still increasing culminated at last in the period of the Great Ice Age.

Glacial Period.

(a) *Boulder-clay.*

The accumulations of the Ice Age must at one time have covered the whole of Cambridgeshire. They still occupy a greater area of it than any other series except the fen, so are of enormous importance to the agriculturist. Being only indifferently exposed and very complex in character they have received but scant attention at the hands of geologists.

Glacial accumulations generally may be divided into two types, Gravel and Boulder-clay, and in Cambridgeshire it is the various modifications of Boulder-clay which are dominant.

The Boulder-clay as we see it is a heterogeneous mass of coarse and fine material containing pebbles and boulders of many sorts of rocks. No stratification is apparent, and the included rock fragments may be of all sizes and shapes, rounded or angular. Soft and compact rocks may have their surfaces smoothed, scratched, striated, and often facetted, and harder rocks, if homogeneous, generally bear evidence of similar treatment.

The constitution of the Boulder-clay varies greatly from place to place, and with changes of constitution there are also great changes of texture.

The constitution of the Boulder-clay at any place is closely related to that of the solid rock beneath it, but is also partly determined by the composition of the rock floor along some definite line, which stretches more or less northward from the place in question.

The matrix of the clay is almost entirely the crushed soft material of rocks near at hand, but the boulders include the survivors of hard and tough rocks from further afield, and the size and abundance of the different boulders is a measure of their resisting qualities and the distance from their point of origin. Large cakes of soft rocks carried from some little distance are also present.

Where the rock underlying the Boulder-clay is moderately hard the junction between the two is very sharp. In other districts it is found that under similar circumstances the surface of the lower rock is grooved, and often polished, but in Cambridgeshire we know of no such 'glaciated pavements.' Sharp junctions of Boulder-clay and Lower Greensand can be seen at Great Gransden, and others are common wherever the clay rests upon Middle or Upper Chalk. In many parts of Cambridge the Boulder-clay rests on soft clays or marls whose upper parts are so disturbed that it is well-nigh impossible to determine any junction.

An exactly similar state of affairs has also been recently described as affecting the Chalk of the highest part of the escarpment near Royston. There, wisps of Boulder-clay underlie parts of the shattered and highly inclined chalk which occurs along the celebrated Royston 'line of flexure,' and Mr H. B. Woodward, who describes the phenomena, refers the production of that 'line of flexure' to the agents which formed the Boulder-clay.

The rock beneath Boulder-clay is generally quite fresh and unweathered, but if soluble, and at the same time porous, it may have undergone considerable alteration by percolating water whose direction of flow the impervious clay above often determines. The rock most easily affected in this way is the Chalk, and between it and the Boulder-clay there are, in some places masses of unweathered flint rubble, while in other places the chalk adjoining the Boulder-clay has been re-cemented and is exceedingly hard.

The general geography of the surface on which the Boulder-clay rests was studied by Mr Jukes-Browne, who concluded that the general form of the country was much the same as at present, but that the chalk escarpment then extended about a mile and a quarter further to the west. He states " that the Cambridge valley, as such, is pre-glacial," and allows that there are certain old valleys, as those of the Lin and Bourn, which were completely choked with Boulder-clay. More recent well-sinkings have shown that filled valleys are more abundant than he supposed, and moreover, that a few of them extend downwards considerably below sea level, e.g. Impington—sixty feet, Saffron Walden and Audley End, but in Cambridgeshire none of these has yet been traced continuously.

Having thus indicated the general relations of the Boulder-clay we may now mention a few of the points of interest presented by the boulders.

The most abundant are those of chalk and flint, which, being ever present, have led to the inclusion of the Cambridge-shire Boulder-clay as a part of the Great Chalky Boulder-clay. The chalk boulders of Boulder-clay which overlies Chalk are often only the local chalk rubble, but those of the clayey accumulations of the west are well rounded and often scratched. They are, moreover, of a type unknown in East Anglia, and strongly resemble certain of the hard beds of the Flamborough chalk. A similar remark applies also to the irregular masses of tabular flints which abound in the west, and these are further interesting in that they are striated on concave as well as on convex surfaces. A few of these foreign chalk and flint boulders occur among the others in the eastern area.

Next in point of numbers come the boulders of the hard members of the Lower Cretaceous and Jurassic series. Many of these are very fossiliferous and lithologically quite characteristic, but much work will have to be done before we are able to refer them to their original homes.

Nodules of Cambridge Greensand which has no known outcrop above the two-hundred-foot contour can frequently be collected from the Boulder-clay right up to the top of the hills about Royston, but do not seem to occur east of Linton. Red Rock like that of Hunstanton is found all over the

county, but is most abundant in the east. Masses of phos-
phate-bearing sandstone like the Spilsby Sandstone and
carrying similar fossils are common in the railway cutting
east of Bartlow, but have not been observed further to the
west. Rounded quartz and quartzite pebbles, red sandstone
and fibrous gypsum, probably Triassic, are common about
Comberton and Old North Road, and have been found to the
east. Friable and porous magnesian limestone has been
collected. Boulders of Carboniferous rocks, especially Carboniferous Limestone and Ganister, are everywhere abundant, and
these are often large. The sandstones are usually rounded
or subangular, but the limestones show most beautiful 'glacia-
tion.' Undoubted Lower Palaeozoic rocks are notable by
their absence, but metamorphic rocks and banded gneisses
are quite common. Quartz-mica schists without felspar,
garnet-mica schists, basic hornblende schists and tourmaline-
quartz rocks are perhaps worth noting, but many types occur.

Of igneous rocks basalt is the commonest, and many
varieties both with and without olivine have been found,
some being exceedingly like parts of the Whin Sill.

Porphyrites probably come next after basalts in point
of abundance, and among them are some like certain Cheviot
rocks. Granites, diorites, rhyolites and a whole host of less
recognizable rocks, some being volcanic, are fairly common.

Last, but not least in interest, we have a series of rocks which
is characterized by containing numerous porphyritic felspars
which in section almost always appear as rhombs. A few of
these are exactly like the famous rocks of the Ringerike
Plateau, west of Christiania, and are certainly Rhomb-por-
phyries, and it is probable that the whole series was once
Norwegian. The commonest of the series is a beautiful granite
porphyry with dihexagonal quartz crystals and pink rhombs of
felspars which weathers to a cream colour. The ferromagne-
sian mineral is not really fresh enough to determine but seems
to have been augite. This rock is already recorded from such
scattered localities as Dullingham, Linton, Royston, Gamlin-

gay, and Impington; and in the district of Old North Road, Bourn, Comberton, Barrington, and Grantchester, which has been explored by the Sedgwick Field Club, it is quite common. Rhomb-porphyry proper is only known from the last mentioned district but is also recorded outside Cambridgeshire in the gravels of Leighton Buzzard and at Hunstanton, so will probably soon be found elsewhere.

Transported cakes of soft rocks do not seem to be particularly common among the Boulder-clay of Cambridge, but the earliest described example was that of the Roslyn pits at Ely. There a mass of undisturbed Gault, Cambridge Greensand and Chalk several hundred yards long occupies an old valley 'ploughed out' of the Kimeridge Clay. At one time Boulder-clay could be seen on all sides of the mass dipping distinctly under it, but now the Lower and Middle Pits are entirely overgrown, and though Gault, Greensand and Chalk can still be observed on the south and east sides of the Great Pit, the relation between the various members is by no means clear.

An even larger cake of Upper Chalk occurs among the Boulder-clay of Catworth, Huntingdonshire, and a well sunk to the Lower Greensand, near Biggleswade, Bedfordshire, has passed through another of Ampthill or Kimeridge Clay. The well section has been recently described by Mr Home, who states that the clay is practically undisturbed and is about sixty-seven feet thick. The included septarian beds dip steeply and the mass is both underlain and overlain by Boulder-clay.

The mass is also interesting as a boulder, for by its fauna and general lithology and by the mode of preservation of its fossils it closely resembles the Lower Kimeridgian beds of Market Rasen (Lincolnshire) and is quite unlike any of the clays of our district.

The exposures of Boulder-clay are generally of the most unsatisfactory description, and in collecting erratics the arable land of the outcrop generally gives a much better yield than Boulder-clay *in situ*.

The only place in Cambridgeshire where Boulder-clay is
worked for economic purposes is the brickyard between Long
Stanton and Over, where a very dark fine-grained and almost
laminated clay with Jurassic fossils and striated chalk boulders
is made into bricks. The more usual exposures are ponds, wells,
ditches, cemeteries, building excavations, and occasional road
or railway cuttings. These last furnish the best exposures,
but as the Boulder-clay is absent from the steeper slopes and
occurs only high on the hills or deep in the valleys few
cuttings are available, and these all where the railways cross
watersheds, as between Bartlow and Haverhill, and about Old
North Road station. A good road section is seen at Hadstock
village. Small streams running over impervious Boulder-clay
sometimes provide exposures, as about Kingston and Bourn,
and chalk or clay pits worked in the underlying solid rock
may also have a little Boulder-clay as at Ely, Shudy Camps,
and Chesterford.

(b) Plateau Gravels.

Above the Boulder-clay of the Chalk Escarpment are the
curious accumulations of rudely stratified material to which
the name Plateau Gravels is often given.

The Plateau Gravels consist, in the main, of the same
materials as does the local Boulder-clay, but contain a far
smaller proportion of chalky material. Their pebbles are
practically unworn and often retain the striae so characteristic
of the boulders in the Boulder-clay, and, in fact, the whole
aspect of the 'gravel' is that of the less soluble residue of the
local Boulder-clay. No indigenous fossils have ever been discovered among the 'gravels,' and the re-sorting of the materials
has been most imperfectly carried out. Plateau Gravels seem
only to occur where the Boulder-clay is exceedingly chalky
and are unknown in the clay districts of the west. They
sometimes appear to overlap the present Boulder-clay and
rest directly on the Chalk. The most accessible exposures are
the higher hill slopes of the spur which joins Barrington Hill,

near Linton, to the Gog Magogs, and the many gravel pits which have been opened at intervals along its southern side.

Similar in many respects to the Plateau Gravels are certain of the 'gravels and loams' which occur at intervals along several of the drift-filled valleys of the Chalk escarpment. The material of these is much more thoroughly sorted than that of the Plateau Gravels, but the bedding is highly contorted and is never continuous over any considerable area. Coarse and fine materials alternate in the most capricious manner, and wisps of what seems to be Boulder-clay come in among them. The loam and many of the finer beds of gravel consist almost entirely of chalk débris (including flints), but the pebbly beds and the 'Boulder-clay' contain abundant erratics and especially ganister. No fossils have yet been discovered and altogether the beds are more like fluvioglacial deposits or Boulder-clay made from older gravels than ordinary river gravels.

The best general exposures are those of the two pits west of Whittlesford station, but the loams are better seen in the railway cuttings south of Chesterford. Similar gravels may also be seen in Wardington Bottom, at the head of the Rhee valley near Royston, and again high up in the Lin valley about Bartlow station. The much-faulted accumulation of loam and gravel seen in a pit about a third of a mile east of Fordham church seems to belong to these, and with it much of the wide-spreading gravels of the Heath country north of Newmarket may also be included.

Post-Glacial and Recent.

The oldest deposits definitely newer than the latest Boulder-clay are the 'gravels of the ancient river system.' These gravels occupy certain now streamless valleys of the chalk escarpment and often cap long ridges which stretch out from the scarp across the low-ground and fen. They differ from the Gravels and Loam series in that they include no wisps of Boulder-clay and are occasionally fossiliferous. Their material includes all constituents of the Boulder-clay and is fairly well sorted though little rolled. False bedding is the rule rather than the exception, and the whole series bears evidence of almost tumultuous accumulation.

The most continuous line of these gravels is that followed by the road from Westley Waterless by Six Mile Bottom, Wilbraham, and Quy to the Barnwell railway bridge at Cambridge. Joining into this at intervals are several 'tributaries,' the most important of which is that from Balsham.

A similar series of gravels caps the ridge which extends from The Observatory to Girton, Oakington, Longstanton, Swavesey, and Over, and again on the north side of the Ouse the line may be further prolonged in the gravel-capped elevations of Bluntisham, Colne, Somersham, Chatteris, Doddington, Wimblington and March. Other less marked lines of gravel come down from the hills about Newmarket, and those gravels of the Heath country which do not belong to the 'Gravel and Loam' series probably belong to this set, as also do some of the Lin valley gravels about Bartlow and Pampisford. Fossils in the landward portions of these gravels are remarkably scarce. Bones of horse, hippopotamus, rhinoceros, and *Elephas antiquus* have been found near Wadham Hall, and Succineas and a few small land-shells may be seen in a gravel pit near Little Wilbraham church. Palaeolithic implements are recorded from various places but are very rare.

The general character of the gravels is well shown in the many gravel pits and cemeteries worked along the outcrop. The most accessible are those along the Newmarket road between Barnwell and Wilbraham, near Girton College, and at Oakington.

About Chatteris occasional marine shells have been found in the gravels, and passing northward these become more and more numerous until at March they are quite abundant. Several small pits have been worked in the neighbourhood of White Lion Lane, about half a mile west of March, and several small sections are now open there. These show some six to eight feet of shell-bearing and false-bedded gravel resting on a very uneven surface of ordinary Boulder-clay which here contains no shells. The shells from the gravel are almost always a good deal worn. The general aspect of the fauna is considerably more arctic than that of the present North Sea. The commonest forms are *Tellina balthica, Turritella terebra, Buccinum undatum, Littorina littorea,* and *Cardium edule,* and with these occur occasional fresh water species including *Cyrena fluminalis.*

The Gravels of the Present River System present less complex problems than do the older ones, but even they exhibit phenomena not yet understood. They occupy the low ground along the larger valleys often to a height of twenty or even fifty feet above the present river level, and they cover nearly all the ground on which Cambridge and the adjoining villages are built.

Three stages, indicated by successive terraces, are distinguishable, and the lower course of the river as thus indicated seems to have varied considerably.

The gravels of the highest and oldest stage are generally spoken of as the Barnwell Series and from their included fossils are the most interesting.

From Barnwell the gravels have been traced along the Cam valley to Chesterford, up the Lin to Linton, by Barton and Comberton to Toft in the Bourn valley, and along the Rhee to Foxton and Barrington. Below Barnwell a wide spread of similar gravel extends across to Histon and one or two patches of it occur further to the north.

Exposures in the Barnwell Terrace are fairly abundant and are often interesting.

At Barnwell the ground is mostly built over but occasional building excavations are made, and when found, they afford a fine collecting ground to the conchologist. *Cyrena fluminalis* abounds in several of the loamy beds, and teeth and bones of large mammalia and especially of Mammoth are quite common.

In the angle between the Milton and Victoria roads at Chesterton several large gravel pits are worked and in these

the character of the gravels is well seen. Lenticles of coarse
and fine material alternate capriciously, and surfaces of local
erosion separate most of the beds. Flints, which form most of
the pebbles, are a little worn but are by no means all rounded.
Some of the finer beds consist almost entirely of re-sorted
chalk, and these, in suffering solution by percolating water,
have often caused contortion of the beds above them. Piping
along ancient tree roots is very common and also complicates
the bedding planes. Small fossils occur only in the marly
lenticles. *Cyrena fluminalis* was at one time abundant but
the mass containing it is now quarried away. Bones of
mammoth, rhinoceros, *Bos primigenius* and deer are often met
with in pockets where they have escaped solution, especially
at the base of the Gravel. The gravel is from fifteen to
twenty feet thick.

Other interesting gravel pits occur at Shelford and Stapleford, also near Sawston and at Comberton, and from all of these
large mammalian bones have at various times been collected.

Much more interesting are the peculiar beds which occur
at Barrington. These are on the whole much finer in texture
than the usual Barnwell gravels and seem to have been
deposited in a lake or backwater behind the Haslingfield chalk
ridge.

The best exposure is that of the above mentioned pits
belonging to the Royston Cement Company. There, resting
in a hollow, eroded in the Chalk Marl, is a variable deposit of
fine and very chalky silt which contains enormous numbers of
slightly worn bones and boulders. The association is truly
remarkable, and it is noticeable that wherever large boulders
are most abundant there too are the best of the bones. The
surface of the Chalk Marl is hard, quite unweathered, and very
pockety, and the best preserved bones always rest directly upon
it. The thickness of the chalky silt may be anything from
a few inches to several feet, and above it come some three
or four feet of ordinary sandy loam with few fossils. A few
pebbles then come in and then another bone-bearing bed which

passes up into ordinary gravelly loam and surface soil. The total thickness is about eight to ten feet. The great majority of the bones from the basal silt are those of Hippopotamus and several good heads of it have been found. Bones and especially teeth of Rhinoceros and the Urus (*Bos primigenius*) are quite common. Mammoth, horse and several species of deer are not rare, and the lion, the hyena, and the bear have been found.

With these are also found large numbers of land and fresh water molluscs of which about ninety species have been recorded. Of these some eighty are still to be found living around Cambridge, but such forms as *Unio littoralis* and *Cyrena fluminalis* are not known nearer than France and Sicily.

The upper bone bed contains only bones of Bos and Bison.

The gravels of the Intermediate Terrace are best seen in the pits at Chesterton. They underlie the greater part of Cambridge town, and extend by Newnham and Brooklands along the west side of the railway to Shelford, where they become inseparable from the later gravels. Similar gravels can also be followed along the Ely road by Milton to Waterbeach and Landbeach where they gradually merge into fen.

The best exposures are those provided by the large gravel pits near the Railway Bridge at Chesterton. The beds are on the whole very like the Barnwell series but their stratification is much more regular, and alternations of coarse and fine material are less frequent. Other good sections may be seen near the Slap Up Inn, close to Waterbeach.

Rolled bones and a few mollusca are occasionally found in both places but are not at all distinctive; *Elephas* and *Rhinoceros* occur.

The Lowest Terrace is the most continuous and the least distinct of all, and rises but little above the modern flood level. Gravels referred to it fringe both sides of all the tributaries of the Cam almost to their sources and sometimes expand to occupy wide areas. Exposures in this terrace are uncommon but the fine gravel which caps several of the Shepreth cement pits seems to belong to this series. Fossils are rare and uninteresting.

Later than all these Gravels is the alluvium of the valleys and the various deposits which form the Fenland. These mark the completion of geological change in Cambridgeshire and show that the rivers had ceased from active erosion and that their lower courses had passed into the condition of water-logged swamps at or about the level of the sea.

The deposits of the Fenland may be divided into three main types, Alluvium, Peat, and Marine Silt, but these from the nature of the case are so mixed up that no boundaries can be drawn between them, and no order of succession is ever maintained.

As might be expected, Alluvium is dominant in the southern part of the area along the river banks; Silt in the northern districts near to the sea; and Peat in all places where the temporary or permanent absence of the other two types has left it free to grow.

The Peat of the Fens is an accumulation of plant remains and includes many species of plants which only grow with roots and stems beneath the water, and indeed it seems probable that it was only when such plants had already prepared a platform that the more usual damp-loving land plants could begin to grow.

Fen Peat, therefore, must have grown outwards quite as much as upwards, so the relative height or depth of particular layers at different localities cannot be any criterion of their relative age and the many small areas which remain as Meres are only areas not yet reached by the outgrowing plants. Whittlesea Mere is a good example.

The best sections of Peat are to be found in areas preserved for Turf-cutting, and of these, Burwell Fen is the most accessible. In the cuttings there, the lower peat consisting of *Juncus* and water plants is generally under water, but the higher layers of damp-loving land plants are very well exposed and are exceptionally free from silt. Experience there has shown that if peat is only removed down to water level it at once begins to grow again and grows at the rate of a foot in twenty years.

In Waterbeach Fen and in the West Fen, near Ely, Turf is occasionally dug, and there among the water-loving land plants, stumps, trunks and branches of forest trees are frequent, and a definite succession of oak and yew, fir, and alder and willow forests has been made out.

About Bottisham, Soham, and in many other parts of the Fens, *Chara* is very abundant in the lower part of the peat, and with remains of molluscan shells provides material for the so-called marls which frequently occur.

The fauna of the peat is much the same as the present fauna of England and nearly all its species are still living. All the mollusca are still living in Cambridgeshire. Of mammals, the only truly extinct species is the great Urus (*Bos primigenius*), but the Roman ox (*Bos longifrons*) has not been seen wild since Roman times. In the peat it is more abundant than any other vertebrate. A head of *Rhinoceros*, a species which is extinct in Europe, was recently found in peaty silt at Little Downham, near Ely, but may be derived from older gravel. The beaver, wolf, brown bear (*Ursus arctos*), and wild boar are now extinct in England but are fairly common in the peat, and with them occur such modern species as the fox, the otter, the marten, the weasel. Among birds the occurrence of the pelican is noteworthy.

The Fen Alluvium and Marine Silt are only exposed in dyke-making, so are but rarely seen. Only included organisms distinguish the fresh-water from the marine deposits, and both are indiscriminately termed 'Buttery Clay' or 'Warp.' They are dark clays, often very carbonaceous and unctuous to the feel. The fossils show that fresh-water and marine conditions have alternated considerably in most of the districts, but on the whole fresh-water conditions have always prevailed south of Littleport.

The fresh-water fauna recorded is meagre and includes only species now common in the district. The marine fauna is notable for its abundant foraminifera which are often numerous enough to render certain beds quite sandy. Bones of the seal, whale, and grampus are occasionally met with. *Scrobicularia piperata* is the commonest and most characteristic mollusc.

VERTEBRATE PALAEONTOLOGY OF CAMBRIDGESHIRE.

By R. LYDEKKER.

FROM the point of view of the student of vertebrate palaeontology Cambridge is an unusually interesting county, since it contains two deposits which are practically unique, although one extends to a certain degree into the adjacent counties. The first of these two deposits is the one at Barrington, yielding mammalian remains of Pleistocene age, remarkable for their perfect state of preservation and for the numbers in which they occur. It is this fine state of preservation and numerical abundance of the remains, coupled with the peculiar nature of the rock in which they are buried, which entitles the Barrington deposit to be called unique, for the species of mammals it contains are not different from those found elsewhere.

The second and more noteworthy fossiliferous deposit is the coprolite band of the Cambridge Greensand, which was so extensively worked for phosphates in the second third of the last century, but is now practically exhausted. From this deposit during the time that the coprolite diggings were in full swing vast quantities of vertebrate remains were secured by various energetic collectors, the greater number of which are preserved in the Cambridge Geological Museum. Unfortunately these remains are for the most part very fragmentary and much rolled and water-worn, so that their determination and association is generally a matter of extreme difficulty. But this is by no means the only unfortunate circumstance connected with these remains. In 1869 was published an

'Index' to the fossil remains of birds and reptiles in the Woodwardian Museum at Cambridge, compiled by Professor H. G. Seeley. Although this was confessedly a mere museum catalogue, generic and specific names were assigned to a large number of the reptilian specimens from the Greensand and other formations of the county, with little or nothing in the way of definition. Some of the forms thus casually named have been subsequently described by the author of the aforesaid 'Index' or by other palaeontologists. Others, however, have remained *in statu quo* to this day, and there is consequently great difficulty in deciding how many of these nominal genera and species should be quoted in palaeontological literature, and how many should be regarded as *nomina nuda.*

Much interest also attaches to the mammalian remains from the Cambridgeshire Fens, although the same types are met with in those of the adjacent counties. A fourth formation in the county which has likewise yielded remains of great interest is the Kimeridge Clay at Roswell pit, near Ely, many of which were collected from about 1850 to 1870 by Mr Fisher, of that city.

Commencing with the remains, from the Fens, the first specimen for notice is a fine skull of the brown bear (*Ursus arctos*) figured on page 77 of Owen's *British Fossil Mammals and Birds* (1846), and described in the two following pages of that work. It is preserved in the Sedgwick Museum, Cambridge, and was dug up in Manea Fen. A second but imperfect bear's skull from the same locality is referred to on page 78 of the work cited. The fore-part of a skull of the otter (*Lutra lutra*), from Littleport Fen, below Ely, is figured in page 119 of Sir R. Owen's work, and is preserved in the Sedgwick Museum.

Jaws of the beaver (*Castor fiber*) were dug up so long ago as the year 1818 about three miles south of Chatteris, in the bed of the old West Water, which once formed a communication between the Ouse and the Nen. They are described by Mr Okes in the *Transactions* of the Cambridge Philosophi-

cal Society for 1822 (vol. I. p. 175)[1]. A skull, without the
lower jaw, from the Cambridgeshire Fens is also figured by
Owen on p. 190 of the work cited above Beaver remains are
also recorded from Burwell, Ditton, Ely, and Watcham Fens[2].

Remains of the great extinct wild ox or aurochs (*Bos
taurus primigenius*) are far from uncommon in the Fens of the
county; some of them bearing evidence as to the existence
of this animal with man. The Sedgwick Museum, for
instance, possesses a skull from Burwell Fen with a polished
(Neolithic) flint implement, or celt, embedded in the forehead[3]; and a second skull from Cottenham Fen, formerly in
the possession of the rector of that parish, likewise displays
evidence of having been injured by a flint weapon. Both
these specimens are figured in Miller and Skertchley's *Fenland, Past and Present* (p. 321). A complete skeleton of the
aurochs from Burwell Fen is preserved in the Zoological
Museum at Cambridge. Apparently side by side with the
wild aurochs lived the domesticated Celtic short-horn (the
so-called *Bos longifrons*), remains of which have been discovered in Swaffham Fen.

Of the deer family the Cambridgeshire Fens have yielded
remains of three species—two still living and the third extinct. Sir R. Owen[4], for instance, considers that part of a
skull with antlers dug up many years ago near Chatteris from
a bed of clay below the peat, belonged to the red deer (*Cervus
elaphus*). This specimen was, however, clearly antecedent to
the true Fen epoch, and indeed it was associated with teeth of
the mammoth. In the fens of various parts of the county red
deer antlers are by no means uncommon. From Burwell Fen
are recorded remains of the great extinct Irish deer or 'Irish
elk' (*Cervus giganteus*), but whether from the peat itself
or the underlying clay does not appear to be certain. Antlers

[1] See Owen, *Brit. Foss. Mamm. and Birds*, p. 195.
[2] See Woodward and Sherborn, *Brit. Foss. Vertebrata*, p. 328.
[3] Described by J. Carter, *Geol. Mag.* (2) I. p. 492 (1874).
[4] *Op. cit.* p. 474.

of the roebuck (*Capreolus capreolus*) from the Cambridge Fens,
where they are comparatively common, are figured on p. 487
of Owen's *British Fossil Mammals and Birds*. Nearly, if
not quite, as abundant in these deposits are remains of the
wild boar (*Sus scrofa ferus*). In the Zoological Museum is
part of the skull of a walrus (*Odobœnus rosmarus*) from the
Fens near Ely. Bird remains from the fens of the county
appear to be rare, or, at all events, have received but little
attention from palaeontologists. The late Prof. Milne-
Edwards[1] however records bones of the coot from these deposits; while Prof. A. Newton[2] has described two specimens
of the humerus (belonging to as many individuals) of a species
of pelican from these deposits. These last specimens, together
with other remains of the same genus described by Dr C. W.
Andrews[3] from prehistoric deposits at Glastonbury, Somerset,
appear to indicate that pelicans bred in England at no very
distant epoch. The wild goose is also recorded.

From the well-known gravel-pits at Barnwell, near Cambridge, as well as from similar deposits in other parts of the
county, have been obtained remains of the usual species of
Pleistocene mammals. The Barnwell list[4] includes the cave
lion (*Felis leo spelaea*), the aurochs, the great Irish deer, the
Pleiostocene hippopotamus (*Hippopotamus amphibius major*),
the wild horse (*Equus caballus fossilis*), the woolly rhinoceros
(*Rhinoceros antiquitatis*), the mammoth (*Elephas primigenius*),
and the straight-tusked elephant (*E. antiquus*). Of the last-
mentioned species an imperfect lower jaw from Whittlesea is
described by Prof. Leith Adams[5]. A mammoth's tusk from
Cambridge, measuring five feet in length, was purchased by
the Royal College of Surgeons in 1842[6]; and Prof. Leith

[1] *Oiseaux Fossiles de la France*, pl. cvi.
[2] *Ibid.* 1868, 363, and *Proc. Zool. Soc. London*, 1871, 702; see also *Dictionary of Birds*, 702.
[3] *Ibid.* 1899, 351.
[4] See Seeley, *Quart. Journ. Geol. Soc.* xxii. 476 (1866).
[5] "British Fossil Elephants" (*Mon. Pal. Soc.*), p. 178.
[6] See Owen, *op. cit.* 248.

Adams records remains of this species from Barton, Chesterton, and Cambridge. The musk-ox is said to occur in the gravel.

The aforesaid mammaliferous deposit at Barrington has been described in considerable detail by the Rev. Osmond Fisher[1]. The first report of the finding of bones in this pit, which is situated in a valley between Haslingfield and Barrington, was brought to Cambridge in 1878; somewhat later a similar deposit was opened up for a short time about half-a-mile higher up the valley. The matrix in which the bones are embedded is a grey sand with a slight admixture of clay. From the fact that many bones of the same animal are often found in their natural association, Mr Fisher is of opinion that the deposits were formed in a deep hole of a stream, where it swept around the sides of the adjacent hill; this stream being none other than the Rhee, which now drains the district. The mammals recorded from Barrington include the cave lion, the cave hyaena (*Hyaena crocuta spelaea*), the badger (*Meles meles*), the cave bear (*Ursus spelaeus*), the aurochs, the Pleistocene bison (*Bos priscus*), the red deer, the great extinct Irish deer, the Pleistocene hippopotamus, the so-called leptorhine rhinoceros (*Rhinoceros leptorhinus*), the mammoth, and the straight-tusked elephant. Very noteworthy is the absence of the woolly rhinoceros, and, more especially, the reindeer (*Rangifer tarandus*), and the great numerical abundance of the remains of the hippopotamus. These latter are remarkable for their great size; specimens of the jaws have been described and figured by Mr P. Lake in the *Geological Magazine* for 1885[2].

A fossil which has given rise to much discussion is a specimen of the conjoint cervical vertebrae of a small cetacean from the Boulder-clay at Ely, preserved in the Sedgwick Museum at Cambridge. In the year 1864 the name *Palaeobalaena sedgwicki* was proposed by Professor Seeley[3]

[1] *Quart. Journ. Geol. Soc.* xxxv. 670 (1879).
[2] Series B, vol. II. 318.
[3] *Proc. Camb. Phil. Soc.* I. 228.

for the cetacean represented by this fossil, but in the following year he changed[1] the title to *Palaeocetus sedgwicki*. It was suggested by its describer that the specimen was derived from the Kimeridge or Oxford Clay. This, however, is an altogether untenable hypothesis, seeing that whales are unknown below strata of Lower Eocene age, and that the forms from these deposits are of an extremely primitive type, whereas the Ely drift specimen conforms in all respects to the corresponding portion of the skeleton of modern whales. A more probable suggestion[2] is that the specimen was derived from the Red Crag of the east coast, and that it belonged to one of the species of beaked whales (*Mesoplodon*) so common in that formation. Although it has been somewhat stained during its entombment in the Boulder-clay, the specimen shows distinct traces of having originally been of the reddish colour characteristic of Crag fossils.

Leaving the superficial deposits, we pass on to those of the Secondary period, of which there is a succession, with some gaps, in the county, from the Oxford Clay to the Upper Chalk. The best chalk-pits in the county are those at Burwell, Cherry-hinton, and Trumpington. Vertebrate fossils do not, how-ever, appear to be very common in these pits; at all events, the published lists of species from the Chalk of the county are meagre. The Lower Chalk at Isleham and the Chalk Marl at Trumpington have yielded some fine jaws and numerous teeth of the great marine reptile *Ichthyosaurus campylodon*, which was one of the last survivors of the group so abund-antly represented in the Lias. This species was established by Mr J. Carter[3], on the evidence of Cambridge specimens. Certain small reptilian vertebrae from Cherryhinton, pre-served in the Sedgwick Museum, have been made the type of a distinct genus and species by Professor Seeley[4],

[1] *Geol. Mag.* II. 54, pl. III.
[2] See Lydekker, *Cat. Foss. Mamm. Brit. Mus.* v. 31.
[3] *Rep. Brit. Assoc.* 1845, 60 (1846).
[4] "Index to Ornithosauria, etc." 3 (1869).

under the name of *Saurospondylus dissimilis.* There is, however, good ground to believe that they are inseparable from the lizard-like marine reptile known as *Dolichosaurus longicollis,* from the Chalk of Kent and elsewhere.

The following species of fishes have been recorded from the Chalk of the county. The dagger-like teeth of the shark known as *Oxyrhina angustidens* are far from uncommon at Cherryhinton, and it is probable that remains of *Lamna appendiculata,* and perhaps of other kinds of sharks, also occur in the same pits. The large crushing palatal teeth of that common Cretaceous ray *Ptychodus decurrens* are likewise met with in the Cambridgeshire Chalk. Among the chimaeroids, the extinct genus *Ischyodus* is represented in the Sedgwick Museum by one half of the lower jaw of the species known as *I. thurmanni,* from the Lower Chalk of Isleham, near Newmarket[1]. Passing on to the bony fishes we find that in the extinct family *Pachycormidae* the well-known Cretaceous fish *Protosphyraena ferox* (often incorrectly called *Saurocephalus lanciformis*) is represented by its dagger-like teeth at Cherryhinton, and more rarely by the solid beak-like extremity of the jaw. Among the more herring-like fishes of the family *Elopidae* we find teeth and jaws of *Pachyrhizodus subulidens* (formerly regarded as those of a lizard) recorded from Cherryhinton, the Sedgwick Museum possessing one particularly fine lower jaw from that locality[2]. The Lower Chalk of Burwell, near Newmarket, has yielded the type specimen of the species known as *Prionolepis angustus,* which is a member of the extinct family *Enchodontidae,* related to the scopeloids and pikes. The type specimen is preserved in the British Museum and its counterpart in the Sedgwick Museum at Cambridge. The species also occurs at Cherryhinton. To the same family belongs *Apateodus striatus,* occurring typically in the Chalk at Lewes, but also recorded from that of Cherryhinton[3].

[1] See Woods' *Type Fossils in Woodward. Mus.* 163.   [2] *Ibid.* 162.
[3] Smith Woodward, "British Chalk Fishes" (*Mon. Pal. Soc.* 1902), 41.

As regards the history of the discovery of vertebrate remains in the so-called coprolite band of the Cambridge Greensand, it may be mentioned that the commencement of the fine collection of bones of pterodactyles in the Sedgwick Museum was made by the Rev. H. G. Day, of St John's College ; other well-known collectors being Messrs Barrett, Carter, Farren, Jesson, and Walker.

By far the most interesting and most important of the vertebrate remains from the Cambridge Greensand are those of birds, the first reference to which appears to have been made by Sir R. Owen[1], who wrote of them as indicating a species of the size of a woodcock. In 1864 the generic names *Pelagornis* (*Pelargornis*) and *Palaeocolymbus* (*Palaeocolyntus*) were proposed by Prof. H. G. Seeley[2] for these birds, but without any definition. The second name was again used by the same writer[3], with a definition, in 1864, but is unavailable on account of preoccupation. It was therefore replaced by the term *Enaliornis*[4], of which two species, *E. barretti* and *E. sedgwicki*, were recognized. When these remains were first discovered they were the only evidence of the existence of birds throughout the strata intervening between the Upper Jurassic horizon of the long-tailed *Archaeopteryx* to the Tertiary, as they still are in this country to the present day. The subsequent discovery of birds furnished with teeth in the Cretaceous of the United States rendered it probable that their approximate contemporaries of the Cambridge Greensand were armed in a similar manner ; this is supported by the fact that the latter, unlike modern birds, have biconcave vertebrae, similar to those of the American Cretaceous genus *Ichthyornis*.

Next in point of interest to those of birds come the

---

[1] *Palaeontology*, 2nd ed. 327, 1861.

[2] *Proc. Camb. Phil. Soc.* I. 228.

[3] *Ann. Mag. Nat. Hist.* (3) xviii. 110.

[4] Seeley, "Index to Ornithosauria, etc. in Woodward. Mus." xvii. (names only), and *Quart. Journ. Geol. Soc.* xxxii. 496 (1876).

remains of flying saurians, or pterodactyles, from the Cambridge Greensand, of which large series are preserved in both the Sedgwick and the British Museums. Some of these bones indicate reptiles of gigantic size. Unfortunately they are all, or nearly all, in a more or less fragmentary condition, which renders the interpretation of their affinities a matter of extreme difficulty and uncertainty. Under these circumstances they are, with one exception, referred to a single generic type, although it is quite probable that, if their condition were more satisfactory, several such groups might be indicated. One of the most remarkable of these bones appears to indicate part of the upper jaw of a toothless pterodactyle allied to *Pteranodon* of the Upper Cretaceous strata of the United States. The form to which it belonged has been named *Ornithostoma*[1], but no specific title has been assigned. If the type specimen be really what it is considered to represent the name *Ornithostoma seeleyi* would be appropriate.

The numerous other forms of pterodactyles from the Cambridge Greensand, here classed in the genus *Ornithochirus*[2] (inclusive of *Cimoliornis*, *Coloborhynchus*, and *Criorhynchus*), have been described by Prof. Seeley[3]. Taking into consideration firstly those represented in the collection of the British Museum, we find in *O. sedgwicki* a large species allied to *O. cuvieri* of the Chalk (which may itself also occur in the Cambridge Greensand), but with a deeper and blunter muzzle and some difference in the form and position of the sockets of the teeth. In *O. fittoni* the beak is much less deep, and the sockets of the teeth are larger and separated by wider intervals. On the other hand, *O. (Criorhynchus) simus* is characterized by the great depth and bluntness of the muzzle, as well as by its huge bodily size. Of species unrepresented in the British Museum *O. denticulatus* is of the approximate size of *O. fittoni*, but has smaller and more numerous teeth ;

[1] See Seeley, *Quart. Journ. Geol. Soc.* **XXXII.** 499 (1870).

[2] See *Cat. Foss. Rept. Brit. Mus.* I. 10.

[3] *Ornithosauria*, Cambridge, 1870.

Owen's *O. daviesi*, from the Folkestone Gault, appears to be closely allied, if not identical. The other Cambridge Greensand forms, of which the bare enumeration must suffice, include *O. brachyrhinus*, *O. capito*, *O. colorhinus*, *O. crassidens*, *O. enchorhynchus*, *O. eurygnathus*, *O. huxleyi*, *O. machaerorhynchus*, *O. microdon*, *O. nasutus*, *O. oweni*, *O. oxyrhinus*, *O. platysomus*, *O. polyodon*, *O. reedi*, *O. scaphorhynchus*, *O. tenuirostris*, and *O. xiphorhynchus*, as well as a few more which have received names but have not been described.

Remains of dinosaurs, or giant land reptiles, are comparatively rare in the Cambridge Greensand, and those that have been found, although in some cases comprising a considerable number of associated bones, are so battered and imperfect that the determination of their true affinities is often impossible. Only two dinosaurian teeth have been recovered from the formation, of which one is noticed below, while the other appears to be lost. Nearly all the Cambridge forms seem to have been of medium size. Some of these dinosaurian remains have been described at considerable length by Prof. H. G. Seeley[1], but of others there is little or nothing in the way of description. Under these circumstances it seems preferable to refer in most cases to the different genera alphabetically rather than attempt to arrange them systematically.

In 1867, Prof. Huxley described from the Chalk-Marl of Folkestone certain dinosaurian teeth and bones, under the new name of *Acanthopholis horrida*. They indicate a member of the armour-clad group belonging to the family *Scelidosauridae*. To the typical Folkestone species certain vertebrae from the Cambridge Greensand have been assigned by Prof. Seeley, who also recognizes a second species from that formation, under the name of *A. eucercus*, other vertebrae and a foot-bone being respectively made the types of species as *A. stereocercus* and *A. platypus*. Possibly, however, the last may be inseparable from the undermentioned *Macrurosaurus*. As to *Anoplo-*

[1] *Quart. Journ. Geol. Soc.* xxxv. 591 (1879).

saurus curtonotus, of which a considerable series of associated remains is known, its describer[1] gives practically no information with regard to its systematic position ; other bones have been named *A. major*. The same may be said of *Eucercosaurus tanyspondylus*[2], which was a reptile of considerable size. *Macrurosaurus semnus* was a smaller type, taking its name from the great length of the tail, the only part definitely known. Another series of tail vertebrae has been described as *Syngonosaurus macrocercus*[3], and appears to indicate a type allied to *Eucercosaurus*. The toothless fragment of an upper jaw described by Prof. Seeley[4] as *Priodontognathus phillipsi*, is regarded as indicating an armoured dinosaur allied to the Wealden *Hylaeosaurus*. Vertebrae in the Sedgwick Museum have been made the type of the genus *Stereosaurus* by the same writer[5], with the species *cratynotus*, *platyomus*, and *stenomus*, but no description has ever been published. Finally, *Trachodon cantabrigiensis* has been established by the present writer[6] on a tooth from the Cambridge Greensand in the British Museum. This tooth indicates a small dinosaur of the iguanodon group, with a pavement-like dentition resembling in general characters the American Cretaceous *Trachodon*, but perhaps generically distinct.

Crocodilian vertebrae in the Sedgwick Museum from the formation under consideration have been described by Prof. Seeley[7] as *Crocodilus cantabrigiensis* and *C. icenicus*, but it may be considered certain that the generic determination requires revision. The head of a femur in the above-mentioned collection is regarded as referable to a lizard by the same writer[7], who has proposed for it the name *Patricosaurus maerocratus*.

[1] Seeley, *op. cit.* 600.
[2] *Ibid.* 612. [3] *Ibid.* 621.
[4] *Ibid.* xxxi. 439 (1875).
[5] "Index to Ornithosauria, etc." xviii.
[6] *Cat. Foss. Rept. Brit. Mus.* i. 244 (1888).
[7] *Quart. Journ. Geol. Soc.* xxx. 693 (1874), and xxxii. 437 (1876).

Chelonian remains are comparatively common in the Cambridge Greensand. Part of a lower jaw in the British Museum has been referred by the present writer[1] to the existing genus *Chelone*, with the name *C. jessoni*; and another specimen of the same part of the skeleton to the Tertiary *Lytoloma*, as *L. cantabrigiense*[2]. Leathery turtles are indicated by *Protostega anglica*[3], assigned to an extinct genus typically from Belgium. A lower jaw in the Sedgwick Museum has been regarded by Prof. Seeley[4] as indicating a land tortoise, and named, without description, *Testudo cantabrigiensis*; but there is little doubt that it does not really belong to the existing genus.

Of much greater interest are the small chelonian skulls from the Cambridge Greensand which have been made the type of the genus *Rhinochelys*[5]. Not improbably these chelonians belong to the *Pleurodira*, a group in which the head is retracted by a lateral flexure of the neck; with some of the existing members of which they agree in having distinct nasal bones. The temporal region of the skull is completely roofed over by bone, as in turtles. The species *R. pulchriceps*, *R. cantabrigiensis*, *R. macrorhina*, *R. elegans*, *R. brachyrhina*, and *R. jessoni* have been defined by the present writer. A number of other specific names have however been assigned by Prof. Seeley[6] to skulls in the Sedgwick Museum, but since no descriptions are given it is unnecessary to quote them *seriatim*. Fragments of chelonian shell from the Cambridge Greensand showing a pustular external sculpture, have been regarded as indicating a distinct genus and species, under the name *Trachydermochelys phlyctaenus*; but there is at present no evidence to indicate that this type of shell does not pertain to *Rhinochelys*[7].

[1] *Cat. Foss. Rept. Brit. Mus.* III. 36 (1889).
[2] *Ibid.* 68.   [3] *Ibid.* 229.
[4] "Index to Ornithosauria, etc." XIX.
[5] See *Cat. Foss. Rept. Brit. Mus.* III. 175.
[6] "Index to Ornithosauria, etc." XVIII.
[7] See *Cat. Foss. Rept. Brit. Mus.* III. 182.

Among the ichthyopterygian, or ichthyosaurian reptiles, *Ichthyosaurus campylodon* of the Chalk is also abundantly represented by teeth and fragments of the jaws in the Cambridge Greensand. Other names have been applied, without definition, by Prof. Seeley[1] to ichthyosaurian remains from the same formation in the Sedgwick Museum. A specimen in the British Museum has been assigned by the present writer[2] to the Jurassic genus *Ophthalmosaurus*, distinguished from *Ichthyosaurus* by the articulation of three (in place of two) bones to the lower extremity of the humerus and femur respectively. The species has been named *O. cantabrigiensis*. The genus is allied to the toothless *Baptanodon* of the North American Cretaceous, but was apparently furnished with teeth in the front of the jaws. Finally, a thigh-bone, or femur, from the Cambridge Greensand in the Sedgwick Museum has been regarded by Prof. Seeley[3] as indicative of a distinct generic type of ichthyosaurian, under the name of *Cetarthrosaurus walkeri*. It is stated "to present a resemblance one degree nearer to the femur of the monotremes than that of *Ichthyosaurus.*"

Teeth and other remains of the second great group of marine Secondary reptiles, the Sauropterygia, or Plesiosauria, are likewise of common occurrence in the Cambridge Greensand, although but few of the species have been satisfactorily defined. Among the most abundant of these are the large deeply-fluted conical teeth of the large-headed and short-necked Cretaceous genus *Polyptychodon*, of which the two species *P. continuus* and *P. interruptus* have been recognized from this formation. Of the more slenderly built and longer-necked types included in the genus *Cimoliosaurus*, specimens referable to *C. bernardi* of the Chalk have been recognized; while several undefined types, such as the so-called *Plesiosaurus planus* and *P. cantabrigiensis* of Seeley, probably belong to the same or a nearly allied genus.

[1] "Index to Ornithosauria, etc." XVII.

[2] *Cat. Foss. Rept. Brit. Mus.* II. 9 (1889).

[3] *Quart. Journ. Geol. Soc.* XXIX. 505 (1873).

Passing on to the fishes of the Cambridge Greensand, it may be noted that teeth of various sharks are among the most common remains in this formation. The more simple types of such teeth have been identified[1] with the species known as *Corax falcatus*, *Lamna appendiculata*, *Oxyrhina angustidens*, *O. macrorhiza*, and *O. mantelli*, most of these ranging upwards to the Chalk, although the fourth is restricted to the Gault and Cambridge Greensand. Of another genus (*Scapanorhynchus*) of the family *Lamnidae* there are three representatives in the Cambridge Greensand, of which *S. gigas* is restricted to that formation, while the other two, *S. subulatus* and *S. rhaphiodon*, are typically from the Chalk. The genus was long supposed to belong to an entirely extinct type, but is now known to be closely allied to, if not identical with, a shark from Japanese waters, named *Mitsikurina*. Fin-spines described as *Spinax major* probably indicate sharks of the genus *Synechodus*, which belongs to the family *Cestraciontidae*, now represented by the Port Jackson shark. Of the comb-toothed sharks (*Notidanidae*) the species *Notidanus microdon*, widely distributed in the Upper Cretaceous, is met with in the formation under consideration.

The solid tooth-like jaws, with their characteristic masticating surfaces, or 'tritors,' of extinct generic types of fishes allied to the living chimaera, or king-of-the-herrings, are far from uncommon in the Cambridge Greensand. Thus the genus *Edaphodon* is represented by the species *E. laminosus*, *E. crassus*, *E. reedi*, and *E. sedgwicki*, of which the second and third are confined to the present formation, while the others have a wide range in the Upper Cretaceous, and are not typically from Cambridgeshire. Of the allied genus *Ischyodus*, the Cambridge Greensand likewise possesses four specific representatives, of which *I. latus* and *I. planus* are confined to that formation, while *I. thurmanni* and *I. incisus* are typically from other horizons.

Among the most beautiful of all vertebrate fossils from the

---

[1] See Smith Woodward, *Proc. Geol. Assoc.* xiii. 196 (1893).

coprolite band of the county are the palatal teeth of various species of pycnodont fishes, all of which are characterized by their crushing type of dentition. Remains of these fishes have been described in two papers contributed to the *Geological Magazine*[1] by Dr A. Smith Woodward. The first genus, *Athrodon*, of which there are two species, *A. crassus* and *A. jessoni*, from the formation under consideration, is known only by the teeth of the splenial element of the lower jaw. These indicate fishes allied to the Jurassic *Mesodon*, but are more irregularly arranged, with the median longitudinal series less well differentiated. The genus *Coelodus*, in which the teeth on the vomer of the palate are arranged in five longitudinal series, while those of the splenial of the lower jaw are elongated transversely and form three rows, is likewise represented by two species peculiar to the Cambridge Greensand. Of these *C. inaequidens* is comparatively common, but *C. cantabrigiensis* is at present known only by a single splenial bone, with its teeth, in the collection of the British Museum. The third genus is the imperfectly known *Anomoeodus*, of which the vomer is nearly flat (instead of convex), and carries teeth of irregular sizes in from three to five longitudinal series, while the splenial dentition is restricted to a small area of the bone, and consists of one large series of teeth, flanked on the inner side by at least one row of smaller teeth, and externally by two or more similar rows. The Cambridge species comprise the large *A. superbus*, the medium-sized *A. confertus*, and the diminutive *A. carteri*. Another type of enamel-scaled, or ganoid fish, the well-known *Lepidotus*, of the family *Semionotidae*, is apparently indicated by certain scales from the Cambridge Greensand in the collection of the British Museum[2]. Another specimen in the same collection indicates a species of the genus *Lophiostomus* peculiar to the Cambridge Greensand, namely *L. affinis*; the genus belonging to the family *Eugnathidae*.

Passing on to fishes of a more modern type, we have

[1] Decade iii. x. 493 (1893), and dec. iv. ii. 207 (1895).
[2] Woodward, *op. cit.* (4) ii. 207.

the problematical *Plethodus expansus* of the Sussex Chalk represented by lower dental plates in the formation under consideration. The lancet-like teeth of *Protosphyraena ferox*, already alluded to when discussing the fishes of the Chalk of the county, are exceedingly abundant in the coprolite beds, as are also the conical beaks of the same fishes. Four other species of the same genus, namely *P. brevirostris, P. depressa, P. keepingi,* and *P. ornata,* have been established on specimens of the last-named element of the skeleton from the Cambridge Greensand.

The Gault of the county appears to be practically devoid of vertebrate remains, the only specimen of that nature with which the writer is acquainted being the arm-bone, or humerus, of a turtle from the Barnwell pits, preserved in the Sedgwick Museum[1].

The coprolite beds of the Lower Greensand of Wicken, Cambridgeshire, have yielded a large number of water-worn bones and teeth of reptiles and fishes similar to those from the corresponding formation of Potton, in Bedfordshire. Most if not all of these remains have been derived from the denudation of preexisting formations. A large series of them is preserved in the Sedgwick Museum. Among the more common of these remains are bones and teeth of *Ichthyosaurus* and of the great short-headed sauropterygians of the Kimeridge Clay known as *Pliosaurus*; some of the Wicken vertebrae having been named by Prof. Seeley, without description, *P. microdirus.* As derivatives from the latter formation may also be reckoned the teeth of the giant crocodilian *Dacosaurus*; while the teeth and bones of *Iguanodon* (the great bipedal herbivorous dinosaur) have been washed out from strata of Wealden age. The scales and button-like teeth of the large Kimeridgian species of the enamel-scaled fish *Lepidotus* are also met with at Wicken, as are likewise the palatal and splenial dental plates of various pycnodont fishes. The teeth and bones from the Wicken deposits, like those from Potton, are stained red by the

[1] Seeley, "Index to Ornithosauria, etc." 73.

ferruginous sand in which they are embedded, and are highly impregnated with phosphate of lime.

The next and last fossiliferous formation of any importance in Cambridgeshire, so far at least as vertebrates are concerned, is the Kimeridge Clay, of which there is a fine exposure in the well-known Roswell pits, near Ely. One of the most interesting reptiles from these pits is the huge marine crocodile known as *Dacosaurus maximus*, the crowns of the teeth of which measure a couple of inches in length. These teeth form smooth compressed cones, carrying a strongly marked vertical keel on each side. The genus and species were first described on the evidence of specimens from the Upper Jurassic of the continent; it is, however, doubtful whether this reptile is really entitled to be separated generically from *Geosaurus*, which is typified by a smaller form from the same horizon. Unlike ordinary crocodiles, *Geosaurus* had an elongated body, a naked skin, and paddle-like limbs, of which the front pair were much larger than the hinder ones. Certain huge vertebrae and limb-bones, inclusive of a claw-phalange, from Ely, in the Sedgwick Museum, indicate a dinosaur allied to *Pelorosaurus* of the Wealden or *Cardiodon* (*Ceteosaurus*) of the Lower Jurassic. For these specimens Professor Seeley[1] proposed the name *Gigantosaurus megalonyx*, but without giving a sufficient description or definition. Not improbably the Ely reptile may be identical with one described on the evidence of a thigh-bone, or femur, from the Kimeridge Clay of Weymouth as *Ceteosaurus humerocristatus*, but which may really belong to *Pelorosaurus*.

Of the two great groups of extinct marine reptiles, the Ichthyopterygia are represented in the Kimeridge Clay of the county by *Ichthyosaurus trigonus*, a species found in the same formation in many other parts of the country, and taking its name from the peculiar shape of the bodies of the vertebrae. Other remains of these reptiles from Ely have been named by Professor Seeley *I. chalarodirus*, *I. hygrodirus*, etc., but have never been definitely described.

[1] Seeley, "Index to Ornithosauria, etc." 73.

Far more abundant than those of Icthyopterygia are the remains of Sauropterygia—and more especially the pliosaur group—in the Kimeridge Clay of the county. Of the typical genus *Pliosaurus*, characterized, among other features, by the large head, short neck, more or less triangular teeth, with one surface nearly smooth and the other two fluted, as well as by the great bodily size of its representatives, at least two species are recorded from this formation in Cambridgeshire. Firstly, there is the typical *Pliosaurus brachydirus*, which is considerably smaller than the undermentioned *P. macromerus*, and has been recorded from both Ely and Cottenham. The typical locality appears to be Oxfordshire. On the other hand, the gigantic *P. macromerus* is typified by a femur from Swindon. In both species the teeth are of the same general type, but whereas there are 35 of these in the lower jaw of *P. brachydirus*, in that of *P. macromerus* there appear to be only 24. Teeth and vertebrae, and less commonly paddle-bones, of the larger species are not unfrequent in the pits near Ely. The name *P. brachyspondylus* has been applied to immature vertebrae of this genus from Ely in the Sedgwick Museum. Nearly allied to *Pliosaurus* is the genus *Peloneustes*, as typified by *P. philarchus* of the Oxford Clay. Among other features, this genus is distinguished by the much greater length of the bony union, or symphysis, of the lower jaw, which includes the sockets of about a dozen teeth. The teeth differ from those of the more typical species of *Pliosaurus* by their nearly conical form and uniform fluting. In the Kimeridge Clay the genus is represented by *P. aequalis*, typified by a femur from Oxfordshire, but also occurring at Ely; the remains from the latter locality having received the name of *Plesiosaurus sterrodirus*[1].

Plesiosaurs seem to be much less common in the Kimeridge Clay of Cambridgeshire than they are in the Oxfordian of Huntingdonshire. The large *Colymbosaurus trochanterius* (at one time included in the Cretaceous genus *Cimoliosaurus*) is, however, represented at Ely. Plesiosaurs of this and the

---

[1] See *Cat. Foss. Rept. Brit. Mus.* II. 153.

allied Upper and Middle Jurassic genera differ from the typical *Plesiosaurus* in the structure of the shoulder-girdle. The present species is specially characterized by the moderate expansion of the lower ends of the humerus and femur, to which three bones are articulated. Another genus of these reptiles is represented in the Ely pits by *Muraenosaurus truncatus*, which is typified by a vertebra from Oxfordshire[1].

Apparently only one type of chelonian has been recorded from the Kimeridge Clay of the county. This species, which belongs to a generalised Jurassic family of the group, was named by Professor Seeley *Enaliochelys chelonia*, but without sufficient description. It was subsequently identified by the present writer[2] with a species from the Kimeridgian of the continent, known as *Thalassemys hugii*. These Upper Jurassic Chelonians had the heart-shaped shells of the modern turtles (of which they were probably the ancestors), but still retained claws on all the digits of the limbs.

Fish-remains appear to be of rare occurrence in the Kimeridge Clay of the county. The head of a species of the genus *Eurycormus*, belonging to the enamel-scaled family *Eugnathidae*, has, however, been obtained from Ely, and is preserved in the Sedgwick Museum. It has been made the type of the species *Eurycormus grandis* by Dr A. Smith Woodward[3], and is the only specimen which fully displays the character of that part of the skeleton of the genus. The species is of large size. The skull and part of the trunk of another enamel-scaled fish in the Sedgwick Museum, obtained from the Kimeridge Clay of Cottenham, were originally described as a new species under the name of *Macropoma substriolatum*, but the generic title was subsequently changed to *Coccoderma*[4]. That genus, although still imperfectly known, is definitely included in the family *Coelacanthidae*.

The vertebrate remains from the Oxford Clay of Cambridgeshire bear no comparison in point of numbers to those

---

[1] *Ibid.* 230.   [2] *Ibid.* III. 148.

[3] *Geol. Mag.* (3) VI. 449, and *Cat. Foss. Fish. Brit. Mus.* III. 354.

[4] See *Cat. Foss. Rept. Brit. Mus.* II. 415.

obtained from the same formation in Huntingdonshire. Probably, however, this is to a great extent due to the circumstance that they have not been collected in the same careful manner as in the latter county. Of pliosaurs the Oxfordian *Pliosaurus ferox*, distinguished from Kimeridgian species by its inferior size and the almost complete absence of facetting on the teeth, which are fluted all round, is not improbably represented by certain vertebrae in the Sedgwick Museum from Great Gransden, for which the name *Pliosaurus pachydirus* has been suggested by Professor Seeley[1]. Remains of the species are also stated to occur in the Oxford Clay of Whittlesea. Of the slender-jawed pliosaurs the abovementioned *Peloneustes philarchus* is represented by the imperfect skeleton of a hind-limb from Whittlesea, preserved in the collection of the British Museum[2]. Among the plesiosaurs, remains of *Muraenosaurus plicatus* and *M. richardsoni* are recorded from Whittlesea in the British Museum *Catalogue of Fossil Reptiles*, where those species are included in the Cretaceous genus *Cimoliosaurus*. In the same work reference is made to the occurrence of bones of an ichthyosaur (*Ichthyosaurus thyreospondylus*) in the clay pits at Whittlesea.

The most interesting specimen from the Oxford Clay of the county is, however, the femur of a small dinosaur, for which the name *Cryptosaurus eumerus* was at first suggested by Professor Seeley. On account of previous use in another sense, the generic title was subsequently changed by the present writer[3] to *Cryptodraco*. The specimen, which is preserved in the Sedgwick Museum, to which collection it was presented in company with a number of other vertebrate fossils, by Mr L. Ewbank, of Clare College, is 12¼ inches in length, and indicates a small member of the iguanodon group.

[1] "Index to Ornithosauria, etc." 18.
[2] *Cat. Foss. Rept. Brit. Mus.* II. 158.
[3] *Quart. Journ. Geol. Soc.* XLV. 46 (1889).

THE MAMMALIA OF CAMBRIDGESHIRE.

By J. Lewis Bonhote, M.A., F.Z.S.

The county of Cambridgeshire cannot be said to have a rich fauna so far as the Mammalia are concerned, and although the number of species actually recorded might compare favourably with those in other counties, yet the number of those species which may be said to be truly indigenous is comparatively small.

The absence of large tracts of wood is probably no small factor in this scarcity; most of the larger animals, such as the Badger, the Marten and the Fox—as well as many species of bats and smaller rodents—being only able to maintain their existence by the shelter and food afforded them in a woodland county.

Roughly speaking, the northern half of Cambridgeshire is flat and woodless, with a few elevations, such as the Isle of Ely, standing up from the surrounding level of the reclaimed Fenland; towards the south and west the county becomes more undulating and wooded, but the district is too small for many rarities to have been found or recorded from it.

The scarcest mammal to which we can lay claim is a bat (*Myotis myotis*), a specimen of which was captured alive about fifteen years ago at Girton. This species, abundant on the Continent, is only known in England, to which it must be considered an extremely scarce straggler, from this and one previous example taken in the British Museum grounds at Bloomsbury. Of other scarce bats, we may note the Barbastelle

(*Synotus barbastellus*) taken at Bottisham on one occasion;
Natterer's Bat (*Myotis nattereri*), a scarce but resident species,
and the Whiskered Bat (*Myotis mystacinus*), which, like the
Barbastelle, owes its place in the county list to a solitary
example. Three species, the Noctule (*Pterygistes noctula*),
the Pipistrelle (*Myotis pipistrellus*), and Daubenton's Bat
(*Myotis daubentoni*) may be considered common and abundant, especially the first two; the last named, however, may
often be seen skimming low over the river near the town, but
does not appear to frequent the open fen country.

Of the Insectivora, all the British species occur in tolerable
abundance; as might be expected, the Water Shrew (*Neomys
fodiens*) is numerous, and, judging from the castings of Owls
and Kestrels, the Pigmy Shrew (*Sorex minutus*) is by no means
rare in the Fens.

Among the Carnivora we have to note the entire absence
of the Wild Cat (*Felis catus*); at the present day it would
naturally not be expected to occur, but there are no records of
any fossil or sub-fossil remains to show its previous occurrence
in the neighbourhood in bygone days. The Badger (*Meles meles*)
can only be considered as a very occasional wanderer, though
possibly still breeding near Wimpole. The Polecat (*Mustela
putorius*) is not very uncommon, especially in the Fen district
and northern half of the county, where one or two examples
are killed yearly; the aquatic habits of this species have
doubtless enabled it to find a suitable home in the Fenland,
where its congener, the Marten, seems to have entirely died
out. The only record of this last-named species is noted by
Jenyns, who says that one was killed at Caxton in 1844, and
according to the same author it used formerly to occur in
Madingley Woods. At a still earlier date, however, it was
probably to be found throughout the county, a subfossil skull
having been taken in Burwell Fen.

The remaining members of this order are by no means rare:
the Fox is scarcer in the open Fenland, to which it is apparently
only a straggler, and the Otter is chiefly to be found in the

river above Cambridge, only occasionally wandering to the more open districts. The Stoat and its near ally, the Weasel, are numerous in all the districts.

It is doubtful whether the Seal (*Phoca vitulina*) should be included in the county list; it is to be found in numbers in the Wash during July and August, and may occasionally wander up the Nene or the Ouse.

As regards the Rodents, all species, with the exception of the Dormouse, occur commonly ; the Dormouse has however only been taken on one occasion near Fulbourn. The Squirrel is abundant wherever the country is wooded, and in the autumn may be found in the more open districts. This county is said to be one of the favourite localities for the little Harvest Mouse (*Mus minutus*), and although evidence of its occurrence during the last few years is wanting, this is probably merely due to lack of observation. De Winton's Field Mouse (*Mus wintoni*) has not yet been recorded, but some very large examples of the Field Mouse (*Mus sylvaticus*) recorded by Jenyns possibly refer to this species.

The old English Black Rat (*Mus rattus*) has not occurred for many years past, but its usurper the Brown Rat, as well as the House Mouse, is abundant.

Coming to the Voles, the Field and Water Vole are numerous, the melanic variety of the latter being by no means uncommon. The Bank Vole (*Evotomys glareolus*) is certainly scarcer than its near ally the Field Vole, but it cannot be considered rare and has been trapped in several distinct localities.

Hares are very numerous everywhere, but more especially in the north-eastern portions of the county towards Newmarket and Mildenhall ; Rabbits are also to be found wherever they are not too rigorously kept down.

The Fallow Deer are now reduced to some 30 head at Chippenham Park ; they used formerly to be kept at Wimpole as well, but the herd at the latter place, then numbering some 300 head, was dispersed about 1880.

The only Cetacean recorded is the Porpoise, which has been known, though rarely, to have ascended both the Nene and the Ouse.

As regards extinct species, the following have been found in the Peat. The Bear, the Walrus, the Beaver, the Wild Boar, the Urus (*Bos primigenius*), the domestic Ox (*Bos longifrons*), the Red Deer, the Roe Deer and the Grampus, a skeleton of the last-named having been found in Thorney Fen. The Red Deer must at one time have been abundant, its remains being among the commonest found, those of the Beaver also being by no means of rare occurrence. Of the Roe Deer, however, only a few antlers have been found, and there are no authentic remains of the Wolf, although it is reported as having occurred. The evidence as to the occurrence of the Reindeer rests on the discovery of a single metatarsal bone, supposed to belong to that animal and figured as such by Prof. Owen.

In the gravel, many remains of an earlier fauna, including some more tropical species, have been found. Among these, the principal are Man, the Bear, the Hyæna, the Lion, the Urus, the Musk Ox, Deer (the three species noted in the Peat), the Irish Elk, the Bison, the Horse, Elephant, Rhinoceros, and Hippopotamus. Most of these specimens have been procured in the gravel beds of Barnwell and Barrington, and from the latter portions of a fine Mammoth were extracted last year.

THE BIRDS OF CAMBRIDGESHIRE.

By A. H. EVANS, M.A., F.Z.S., Clare College.

I. INTRODUCTORY.

CAMBRIDGE and the districts in the vicinity have always been of the greatest interest to ornithologists, owing to the proximity of the Fen country, of old the haunt of many species of Birds which were to be found breeding in few other parts of Britain. The Marsh-Harrier, the Bittern, the Great Bustard, the Ruff and his consort the Reeve, the Black-tailed Godwit, and the Black Tern will doubtless first present themselves to the mind of the reader, as having either entirely or almost entirely ceased to rear their young in the kingdom; but even more remarkable is the case of Savi's Warbler, which was only recognized as a regular summer visitor to the Eastern Counties early in the last century, and disappeared finally from the country in 1856. Bones of Pelican[1], Swan and Wild Goose have been found in the peat in company with those of commoner species, but with nothing to indicate the exact period to which they belong.

Owing, however, to the gradual drainage of the Fens and

[1] The first bone of a Pelican recognized in this county and country was the *humerus* of a bird so young as not to have the *epiphyses* attached, so that the inference of its having been bred in the district is almost certain. The species is no doubt *Pelecanus crispus*, as its large size shows, and other bones of the same species have been found at almost precisely the same place, the evidence being to the effect that they belonged to at least three individuals.

the consequent extension of cultivation to large areas formerly occupied chiefly by sallow-bushes, reed-beds and sedges, the state of affairs has entirely changed since the beginning of the nineteenth century, and Cambridgeshire can now lay claim to but little of her ancient glory as a paradise for birds of the moor and morass. Wicken Sedge Fen and the surrounding grass lands still provide breeding quarters for an occasional pair of Montagu's Harriers or of Short-eared Owls, but instances such as these only serve to emphasize to the student of the past the great changes that have taken place in the condition of the country and their effect upon its feathered inhabitants.

Before entering upon a detailed account of the species of Birds now or formerly found in the county, it may be well to state briefly the character of the immediate neighbourhood and its surroundings, which has so decided an effect upon their presence and numbers[1].

The great 'Bedford Level,' containing the whole of the ancient Fens, and including the Cam valley gault tract, occupies all Cambridgeshire north of an imaginary line from about Newmarket to Huntingdon; here large woods are conspicuous by their absence, and the trees, chiefly Black Poplars and Willows, either form long shelter strips only a few yards wide, with occasional isolated clumps, or stand in rows along the side of some watercourse; the vegetation in uncultivated spots is tall and rank, rushes are a common feature of the landscape, and wide 'lodes' of water connected by narrower channels constantly take the place of walls or hedges.

Southward of this tract comes an area composed of gault, chalk and chalk marl, topped by gravel or to the west by clay, which extends from the Fen districts to another imaginary line from near Newmarket to Biggleswade; here are woods of considerable size, consisting of hard-wood trees, mingled in places with larch or fir, and the general appearance is that of many of our Midland counties.

[1] A fuller account will be found in the Introduction to Babington's *Flora of Cambridgeshire*.

Finally, to the south of this the chalk lands rise gradually as plateaux or hills of moderate height, characterized by plantations of beech and fir, and by the usual vegetation of dry down-lands. In addition to these main divisions the parish of Gamlingay on the extreme west, and the Isle of Ely proper, are chiefly of the Lower Greensand formation ; at Chippenham an expanse of loose sand and gravel overlying the chalk leads to the similar country towards Brandon and Thetford in Norfolk, while at the extreme north of the county, above and below the town of Wisbech, is a dried-up marsh, formed by the silt of the river Nene. Isolated eminences or ridges of Boulder-clay or gravel are also scattered over the Fen districts.

No account is here taken of the state of the country before the period of the Roman occupation and for some time subsequently, when it is supposed to have been kept fairly dry by natural drainage, and was covered in many places by woods, orchards, or vineyards ; for even then it is evident from the bones discovered in the Fens that a certain amount of marsh land must have existed.

No large sheets of water are now to be found in the county[1], and even Whittlesey Mere, just within the Huntingdonshire boundary, was drained in 1851, while, as might be expected from the fact that the town of Cambridge is little more than from twelve to fifteen feet above the average tide level in Lynn Deeps, the rivers run sluggishly between their muddy banks.

The Avifauna of the different areas naturally depends to a great extent upon their physical characters : in the Fens, for instance, the Sedge, Reed and Grasshopper Warblers, the Reed Bunting and the Snipe, with an occasional Montagu's Harrier or Short-eared Owl, as above-mentioned, are the most characteristic species ; in the gravel and clay district the

---

[1] Soham Mere must once have been of considerable dimensions, as is shown by ancient maps, while Stretham Mere was a similar but smaller piece of water. The former was drained by 1793 and the latter even earlier.

majority of the Warblers—among which the Nightingale is
extraordinarily plentiful—with other woodland birds, and
Woodpeckers to the westward, are perhaps the most noticeable; on the high chalk lands the Mistletoe Thrush, our
winter guest the Brambling, and the Long-eared Owl, have
their head-quarters, while the Stone-Curlew is still known to
occur in a few places in the breeding season. Ducks, Waders,
and Sea-birds are for the most part conspicuous by their absence
in summer.

II. Some Birds which have been exterminated from
       the county, or are now extremely rare.

1.   *Locustella lusciniöides* (Savi).   Savi's Warbler[1].

The history of the discovery of this summer visitor to
England is of exceptional interest. Unnoticed by ornithologists until about the year 1819, and only recognized as distinct
from its nearest allies by Savi in 1824 from specimens obtained in Tuscany, there can be little doubt that it had long
been known—though probably mainly by its note—to many
of the marshmen, under the name of 'Brown-,' 'Red-,' or
'Night-Reeler' as a different bird from the Grasshopper
Warbler, or 'Reeler' proper. The first example brought to
the notice of naturalists was shot in the month of May at
Limpenhoe in Norfolk, by Mr James Brown of Norwich, early
in last century, and was submitted to Temminck, who happened
to be in London in 1819. Unfortunately he mistook it for
a variety of the Reed Warbler, and, subsequently as it seems,
for Cetti's Warbler, so that when Mr G. R. Gray in 1840
received two specimens from Mr Baker, of Melbourn in Cambridgeshire—said, but perhaps wrongly, to have been procured
near Duxford—he simply referred them to Savi's new species,
in ignorance of the previous discovery. From that time

[1] The classification and nomenclature adopted are those of Mr Howard
Saunders' *Manual of British Birds*, 2nd ed.

onwards specimens were procured in Norfolk, in Huntingdonshire, and in this county, especially in the parish of
Milton, where certain curious old nests, entirely composed of
the broad leaves of *Glyceria aquatica*, had been the cause
of frequent astonishment to the sedge-cutters[1]. These men
finally discovered in 1845 a fresh nest with eggs, which were
purchased by Mr Bond and by him distributed to various
ornithologists. That gentleman obtained in all two sets of
eggs and six birds. Many other nests were afterwards found
in the three counties already named, but the reclamation of
the Fens seems to have banished the bird from Cambridgeshire
by 1849, and the last specimen known to have been obtained
in Britain was shot at Surlingham, in Norfolk, in 1856.

2. *Panurus biarmicus* (Linn.). Bearded Titmouse.

This beautiful little bird, known to the Norfolk marshmen
as the 'Reed Pheasant,' and still maintaining itself, though
in diminished numbers, on the 'Broads,' was formerly in the
habit of resorting to certain places in the Fens, usually after
the breeding season was past[2]; but there is no certain evidence
that its nest, delicately woven with sedges and the like and
lined with the flowering tops of reeds, was ever actually found
within our boundaries. It was, however, a resident in considerable numbers at Whittlesey and Ramsey Meres in Huntingdonshire, where the reed-beds were sufficiently large to suit
its requirements, while Soham Mere of old would have been
the most likely locality in Cambridgeshire.

3. *Asio accipitrinus* (Pallas). Short-eared Owl.

This species is one of the very few rare British birds which
still visit Cambridgeshire for the purpose of reproduction. It

[1] One specimen of the bird killed about this time is in the University
Museum of Zoology, as are also one or two nests.

[2] Mr J. H. Gurney (*Trans. Norfolk and Norw. Nat. Soc.* VI. p. 437)
mentions the bird as seen at Roswell Pits, near Ely, in 1897 and 1898.

breeds or attempts to breed at intervals in or near Wicken
Fen,—an area now strictly preserved by the County Council,—
and possibly elsewhere, though no absolute case is on record
for other districts of late years. Mig.atory individuals are,
moreover, often met with by sportsmen in autumn, but the
bird's visits are quite irregular, for, while it may not be
observed in summer for several years in succession, it may
suddenly occur in numbers when circumstances are suitable.
A well-known case was during the 'vole-plague' of 1890–91,
on the Scottish border-lands.

4.   *Circus cineraceus* (Montagu).   Montagu's Harrier.

This scarce bird-of-prey still visits the neighbourhood of
Wicken Fen with some regularity. Owing to the fact of its
absence during the colder months, and the lack of game-
keepers in the Fens, it has a better chance of survival than
other members of its family, but it is not observed every
year in the county, and when observed does not invariably
breed with us. The food consists largely of reptiles, amphibians and insects, and, as in the case of Owls, any harm
that is done is much more than compensated by the destruction of noxious creatures.

5.   *Ardetta minuta* (Linn.).   Little Bittern.

There can be little doubt that this species has nested of
comparatively recent years on the Norfolk Broads, but as
regards Cambridgeshire the fact is less certain. Still it is at
least probable that a pair which were shot near Ely in the
spring of 184.., of which the male is still preserved in the
University Museum, were intending to breed with us. We
may therefore fairly infer that the Little Bittern was a more
or less regular migrant to our shores before the drainage of the
Fens, while the later instances of its occurrence give evidence
of a tendency to resume its former habits.

6. *Botaurus stellaris* (Linn.). Common Bittern.

Considerable numbers of the Common Bittern certainly bred in this county in the eighteenth and up to the beginning of the nineteenth century, a nest having been found near the Cam so late as 1821. Probably it was once a more common bird than is usually supposed in Britain, for records of such matters were but seldom kept in earlier times, but since 1868 no nest has been actually discovered, although the booming note has been heard in Norfolk in the spring season within the last twenty years, and no doubt an occasional pair would breed in suitable localities if unmolested. Drainage operations have apparently been the main cause of the disappearance of this migratory bird, but there is some slight hope of its resuscitation now that adequate protection is likely to be afforded. It may be remarked that bones of the Bittern occur in some numbers in the peat deposits.

7. *Anser cinereus* (Meyer). Grey-Lag Goose.

This species is doubly interesting to those who study the ornithology of Cambridgeshire, for not only have bones attributed to it been found in the peat of the Fen districts, but there is also direct evidence that it bred within our limits until 1773. In these latter days, when—as far as Britain is concerned—it is confined as a resident species to the north of Scotland, it seems almost impossible to believe that it could have been common in the eastern counties of England little more than a century ago; but such was undoubtedly the case, and it is equally certain that no other Goose has yet been proved to have nested in a wild state in the kingdom. Daniel, in his *Rural Sports*, Pt II. Vol. II. p. 244 (1802 ?), says :— "This species inhabits the English fens, and it is believed does not migrate, as in many countries on the Continent, but resides and breeds in the fens : they sit thirty days, and hatch eight or nine young, which are often taken; are esteemed most excellent meat, and are easily made tame.

The compiler took two broods in one season, which he turned
down, after having pinioned them, with the common geese."

8. *Porzana maruetta* (Leach). Spotted Crake.

Though the Water Rail, of which an occasional pair may
still be found breeding in this county, occurs in winter as well
as in summer, its near relation the Spotted Crake is a migrant
from abroad. This of itself would make it more difficult
to define its *status*, while the bird's skulking nature often
prevents it from being observed. On the other hand, if an
individual is detected in spring, it is probable that it intends
to remain for the summer, and we may hope that future in-
vestigations will show that this species has not entirely ceased
to breed with us, for it seems to have been common, at least
in some years, up to about 1850, and it does not require so
large an extent of sedge as do some other marsh birds.

9. *Porzana bailloni* (Vieillot). Baillon's Crake.

In the *Zoologist* for 1859 (p. 6329), and again in Gould's
Birds of Great Britain, will be found an account of the discovery of two nests of this Crake in Cambridgeshire, in June
and August 1858 respectively, some of the eggs being now in
the Wolley collection at the University Museum of Zoology.
The breeding of this species in England was corroborated by the
further discovery of two other nests attributed to it, in 1866,
on the Norfolk Broads; but these instances appear to have
been quite exceptional. The somewhat similar Little Crake
(*P. parva*) has not yet been proved to breed in Britain.

10. *Grus communis* (Bechstein). Crane.

Dr William Turner, writing in 1544, says of this bird,
"Apud Anglos etiam nidulantur grues in locis palustribus,
et earum pipiones saepissime vidi," which, if it cannot be
taken precisely as evidence that the Crane bred at that
time in Cambridgeshire, makes it at least somewhat more
than probable, as Turner was a Cambridge man, who held

office for several years at Pembroke Hall. It is moreover
extremely unlikely that it should not have bred in the
Fens, if it did so in the adjoining counties, the only other
interpretation to be put upon Turner's words. About fifty
years later Dr Thomas Muffet corroborates the fact to some
extent (cf. Yarrell, *Hist. Brit. Birds*, ed. IV. p. 180), and
indeed it is perfectly certain that the bird was quite common
in early times in Britain, when it was a frequent quarry for a
Falcon. King John, for instance, procured seven individuals at
Ashwell in this county in December 1212. The Crane occurred
at all seasons of the year, and was met with in large flocks
until after 1678, though at that date Willughby tells us
that he was unable to record an instance of its nesting. The
exact date of its final disappearance as a breeding species,
and even as an abundant immigrant, is quite uncertain;
but Pennant in 1768 concluded that it had forsaken Britain.
Bones have been occasionally found in the Cambridgeshire
peat and at Lynn in Norfolk.

11. *Otis tarda* (Linn.). Great Bustard.

This fine bird, which was an abundant resident in many
parts of England until the beginning of last century, and
used at one time to occur also in the south of Scotland, was
probably more plentiful in the eastern counties than elsewhere, and so continued until about 1812, though exact
dates cannot be given, and but few individuals remained
by that time in other parts of England. The last nests,
in Norfolk and Suffolk, were found in 1832, but solitary
hens seem to have dropped an occasional egg at random
for some six years later. The use of the horse-hoe for weeding
purposes after the corn began to be sown in drills is said to
have been the final cause of extermination in Norfolk, but the
cultivation of lands that were formerly waste, and the extension
of plantations, must have greatly reduced the numbers of
the birds there and elsewhere. Newmarket and Royston
Heaths were well-known resorts of the Bustard in olden times,

but they have long ago ceased to provide those quiet haunts which to this shy bird seem indispensable. The eggs were protected by a law of Henry VIII. in 1534, under the same penalty as those of the Crane—twenty pence apiece, but in neither case, as will be seen above, does the Act seem to have produced the desired effect. Attempts have been made of late years to re-establish the Bustard at Feltwell, near Brandon, and at Elveden, near Thetford, but at present without success. A few individuals, however, still occasionally stray to Britain.

Professor Newton has kindly furnished me with the following additional notes on this species :—

"Two stuffed examples of the Bustard from Cambridgeshire are in the University Museum, but whence they were procured is not known: either or both may possibly be of foreign origin, and not of the old English breed.

"The late Mr Joseph Clark of Saffron Walden told me that it was said that one of them was shot by a man of the name of Davy of Hinxton, his gun being loaded with a black-lead pencil. The bird was wounded, but flew to Shelford, where I suppose it was taken. The story of the other specimen is that it was killed at Ickleton.

"The Museum has also two other English Bustards, one from Icklingham in Suffolk. Of the second nothing is known.

"The Museum contains an egg taken in Cambridgeshire, and given to the Philosophical Society in 1831 by a Mr Barron —concerning whom I have never been able to make out anything.

"The late Dr William Clark, Professor of Anatomy, used to say that once, while riding over Trumpington Heath, he came upon a strange bird on the ground. He took it up and brought it to Cambridge, when, finding that it was a young Bustard, he rode back to the place where he found it and left it there.

"At the end of February and beginning of March, 1856, a bustard frequented Burwell Fen and the neighbourhood for

several weeks. Mr Frederick Godman and the late Mr Anthony
Hamond saw it more than once. The people would not let it
alone, and it was shot at several times. My brother Edward,
Mr Salvin and I, went to the place, but the bird had left.
We, however, saw its footmarks in a field of cole-seed, and
found some feathers—one of which I now have."

12. *Machetes pugnax* (Linn.). Ruff and Reeve.

The Ruff, and its consort the Reeve, were abundant in the
marshy districts of the east of England until a comparatively
recent date, and were by no means unknown in suitable places
elsewhere south of the Border. The diminution in the number
of the birds breeding with us has been so gradual that little
can be said except that they have become much scarcer since
the beginning of the nineteenth century owing to improved
systems of drainage, but an occasional nest has been found
in Norfolk or Lincolnshire almost up to the present time, and
it is by no means certain that one or two pairs do not still
breed annually in the former county. The 'hilling' of the
polygamous males, that is, their practice of congregating on
dry hillocks in the marshes to spar bill to bill with expanded
ruffs and ruffled feathers for the possession of the females,
has been so well described by Montagu and other writers on
British birds that it needs no repetition here; while the
method of netting both sexes, with or without the use of
decoys, is equally familiar to most readers. Large numbers
of these migratory birds used to be taken annually and
fattened for the market on bread and milk or boiled wheat.

13. *Limosa belgica* (J. F. Gmelin). Black-tailed Godwit.

Though much rarer than the preceding species this Godwit
used to breed in certain favoured localities in Cambridgeshire,
Yorkshire, Lincolnshire, and Norfolk until the beginning of
the nineteenth century; but, according to the Rev. L. Jenyns,
it had become scarce by 1825, while the last nests appear to

have been found in our county about 1829, and eighteen years later in Norfolk. Always a migrant the bird still continues to visit our shores in spring and autumn, but is hardly likely, in view of the change in the condition of the country, to remain with us again during the summer. An egg, now in the University Museum of Zoology, was bought in the Cambridge market in 1847.

14.　*Hydrochelidon nigra* (Linn.).　Black Tern.

This species, once a regular migrant to England for purposes of reproduction, and by no means uncommon on many of our fens and marshes, has long since ceased to breed there, the last eggs having been found in Norfolk in 1858. The 'Blue Darr,' or 'Car-Swallow,' as it is locally named, may often be seen as late as, or even later than May, round the Broads or the smaller waters of the eastern counties, but no evidence of its nesting of recent years is obtainable, while in Cambridgeshire it probably was never particularly common, though immense flocks appeared near Bottisham in 1824, and at least one nest was discovered. In 1831 flocks were seen at Gamlingay.

For further details of the above-mentioned species, or those in the list below, the reader may be referred to

Yarrell's *History of British Birds*, ed. IV.

Hewitson's *Coloured Illustrations of the Eggs of British Birds.*

Professor Newton's *Dictionary of Birds.*

Mr Howard Saunders' *Illustrated Manual of British Birds.*

And for the county in particular

Jenyns' *Observations on the Ornithology of Cambridgeshire,* in the second volume of the *Transactions of the Cambridge Philosophical Society.*

The Victoria History of the Counties of England. Cambridgeshire. (Article on Birds by the present writer.)

III. Full list of Species recorded from
Cambridgeshire[1].

*[Square brackets imply that the species is recorded doubtfully
from the county.]*

Turdus viscivorus, Linn. Mistletoe-Thrush. Plentiful.

T. musicus, Linn. Song-Thrush. Common.

T. iliacus, Linn. Redwing. A fairly abundant winter visitor.

T. pilaris, Linn. Fieldfare. „ „ „ „

T. merula, Linn. Blackbird. Common.

T. torquatus, Linn. Ring-ousel. Occurs rarely on the
autumn migration, and even less commonly in spring.

Saxicola oenanthe (Linn.). Occasionally observed in various
places on migration, and breeding in a few.

Pratincola rubetra (Linn.). Whinchat. A fairly common
summer migrant.

P. rubicola (Linn.). Stonechat. Resident, but very local.

Ruticilla phoenicurus (Linn.). Redstart. A moderately
common summer migrant.

Erithacus rubecula (Linn.). Redbreast. Common.

Daulias luscinia (Linn.). Nightingale. An abundant summer
migrant in most parts.

Sylvia cinerea, Bechstein. Whitethroat. A common summer
migrant.

S. curruca (Linn.). Lesser Whitethroat. A summer migrant,
more plentiful with us than in most districts.

S. atricapilla (Linn.). Blackcap. A plentiful summer migrant.

S. hortensis, Bechstein. Garden-Warbler. „ „

S. nisoria (Bechstein). Barred Warbler. Has once occurred
in Cambridge (*Proc. Zool. Soc.*, 1879, p. 219).

S. undata (Boddaert). Dartford Warbler. A rare straggler.

Where not otherwise mentioned a species may be considered
resident, though many individuals doubtless leave Britain for the
winter, their places being taken by accessions from abroad. In some
cases only a few of the birds remain at that season.

Regulus cristatus, K. L. Koch.   Golden-Crested Wren.   Not
    very common.
[*R. ignicapillus* (C. L. Brehm).   Fire-Crested Wren.   The
    original record for Britain, founded on a bird taken
    near Cambridge in 1832, is considered more than doubt-
    ful.   A fresh record is needed.   (*Proc. Zool. Soc.*, 1832,
    p. 139; *Zoologist*, 1889, p. 172.)]
Phylloscopus rufus (Bechstein).   Chiffchaff.   A widely dis-
    tributed summer migrant, nowhere very abundant.
P. trochilus (Linn.).   Willow-Wren.   A moderately common
    summer migrant.
P. sibilatrix (Bechstein).   Wood-Wren.   Has been reported
    from Cambridge and Bottisham, a specimen from the latter
    place being in the University Museum of Zoology.
Acrocephalus streperus (Vieillot).   Reed-Warbler.   A local
    summer migrant, plentiful where it occurs.
A. palustris (Bechstein).   Marsh-Warbler.   The nest is stated
    to have been taken in Cambridgeshire (see Saunders'
    Manual, ed. 2, p. 82), but there is no other record for
    the county.
A. phragmitis (Bechstein).   Sedge-Warbler.   A very common
    summer migrant.
Locustella naevia (Boddaert).   Grasshopper-Warbler.   Plenti-
    ful in Wicken Fen, an uncommon summer migrant else-
    where.
L. luscinioïdes (Savi).   Savi's Warbler.   See above.
Accentor modularis (Linn.).   Hedge-Sparrow.   Common.
A. collaris (Scopoli).   Alpine Accentor.   The first record for
    Britain was that of two specimens at King's College,
    Cambridge, in 1822.   None have been observed in the
    county since that year.   (*Zool. Journ.* I. (1824), p. 134;
    Yarrell, ed. iv. I. p. 291.)
Panurus biarmicus (Linn.).   Bearded Titmouse.   See above.
Acredula caudata (Linn.).   Long-tailed Titmouse.   The form
    separated as *A. rosea* is that common in Britain; *A. cau-
    data* proper has not occurred in Cambridgeshire.   This

species is often seen after the breeding season in flocks, which probably consist to a great extent of visitors to the county, as the nest is not very frequently discovered there.

Parus major, Linn. Great Titmouse. Common.

P. ater, Linn. Coal-Titmouse. The British form (*P. britannicus* of Sharpe and Dresser) is moderately common.

P. palustris, Linn. Marsh-Titmouse. Local and by no means common.

P. caeruleus, Linn. Blue Titmouse. Common.

Sitta caesia, Wolf. Nuthatch. Local and nowhere abundant.

Troglodytes parvulus, K. L. Koch. Wren. Common.

Certhia familiaris, Linn. Tree-Creeper. Fairly plentiful.

Motacilla lugubris, Temm. Pied Wagtail. Abundant.

M. alba, Linn. White Wagtail. This species must be considered a visitor to the county on migration, until further proof of its supposed nesting with us is obtained. It occasionally interbreeds with the Pied Wagtail.

M. melanope, Pallas. Grey Wagtail. A rare winter visitor.

M. raii (Bonaparte). Yellow Wagtail. A very common summer visitor.

Anthus trivialis (Linn.). Tree-Pipit. A local summer migrant, nowhere very plentiful.

A. pratensis (Linn.). Meadow-Pipit. Not uncommon, but somewhat local.

Lanius excubitor, Linn. Great Grey Shrike. A scarce winter visitor.

L. collurio, Linn. Red-backed Shrike. A plentiful summer visitor.

L. pomeranus, Sparrman. Woodchat Shrike. Has been obtained once, near Swaffham Prior, before 1840. The specimen is in the Saffron Walden Museum.

Ampelis garrulus, Linn. Waxwing. Occurs at intervals in winter, singly or in small flocks.

Muscicapa grisola, Linn. Spotted Fly-catcher. An abundant summer visitor.

M. atricapilla, Linn. Pied Fly-catcher. Has occurred in
May, as at Hinxton in 1836. (Saffron Walden Museum.)

Hirundo rustica, Linn. Swallow. Abundant. A summer
visitor.

Chelidon urbica (Linn.). Martin. Plentiful. A summer visitor.

Cotile riparia (Linn.). Sand-Martin. A common summer
visitor.

Ligurinus chloris (Linn.). Greenfinch. Common.

Coccothraustes vulgaris, Pallas. Hawfinch. Rare and local.

Carduelis elegans, Stephens. Goldfinch. Not uncommon.

C. spinus (Linn.). Siskin. An occasional winter visitor,
sometimes observed in large flocks[1].

Passer domesticus (Linn.). House-Sparrow. Very common.

P. montanus (Linn.). Tree-Sparrow. Abundant.

Fringilla coelebs, Linn. Chaffinch. Very common.

F. montifringilla, Linn. Brambling. Not uncommon in
winter in many places. Large flocks are often met with.

Linota cannabina (Linn.). Linnet. Fairly common, but
somewhat local.

L. linaria (Linn.). Mealy Redpoll. A specimen was obtained
in May 1836 at Hinxton. (Saffron Walden Museum.)

L. rufescens (Vieillot). Lesser Redpoll. Scarce.

L. flavirostris (Linn.). Twite. Strays to the county in
autumn, and is sometimes seen in large flocks.

Pyrrhula europaea, Vieillot. Bullfinch. Fairly common.

P. enucleator (Linn.). Pine-Grosbeak. Has once been obtained in the county, a specimen, now in the Saffron
Walden Museum, having been shot by the groom of
the Rev. A. H. D. Hutton in the garden of Little
Abington vicarage on January 13, 1882 (*Zoologist*, 1883,
p. 222).

Loxia curvirostra, Linn. Crossbill. A rare straggler, singly
or in flocks.

Emberiza miliaria, Linn. Corn-Bunting. Abundant.

---

[1] Turner (*Avium praecipuarum—historia*, 1544, p. 88) mentions having
seen the Siskin in Cambridgeshire.

E. citrinella, Linn. Yellow Bunting or Yellow-hammer. Fairly common.

E. cirlus, Linn. Cirl Bunting. Very rare.

E. schoeniclus, Linn. Reed Bunting. Local.

Calcarius lapponicus (Linn.). Lapland Bunting. A rare straggler in certain winters, as in 1826 and 1892.

Plectrophenax nivalis (Linn.). Snow-Bunting. An uncommon winter visitor.

Sturnus vulgaris, Linn. Starling. Common.

Pastor roseus (Linn.). Rose-coloured Starling. A very rare summer visitor.

Nucifraga caryocatactes (Linn.). Nutcracker. A rare straggler in winter; an example killed near Wisbech on November 8, 1859, proved to belong to the thin-billed race.

Garrulus glandarius (Linn.). Jay. Not uncommon.

Pica rustica (Scopoli). Magpie. Has become rare.

Corvus monedula, Linn. Jackdaw. Common.

C. corax, Linn. Raven. Used to breed in the county, but is now exterminated.

C. corone, Linn. Carrion-Crow. Not uncommon.

C. cornix, Linn. Hooded Crow. Plentiful in winter.

C. frugilegus, Linn. Rook. Abundant.

Alauda arvensis, Linn. Sky-lark. Very common.

A. brachydactyla, Leisler. Short-toed Lark. Has occurred once in autumn, a specimen having been procured by a bird-catcher near Cambridge. (*Zoologist,* 1883, p. 33.)

Cypselus apus (Linn.). Swift. A plentiful summer visitor.

C. melba (Linn.). Alpine Swift. Jenyns (*Fauna Cantabrigiensis* MS. in the University Museum of Zoology) records a specimen killed between Cambridge and Grantchester in September, 1838.

Caprimulgus europaeus, Linn. Nightjar or Goatsucker. A summer migrant, local, and nowhere very common.

Iynx torquilla, Linn. Wryneck or Cuckoo's Mate. Now a somewhat rare summer visitor.

Gecinus viridis (Linn.). Green Woodpecker. Not uncommon and in some districts abundant.

Dendrocopus major (Linn.). Great Spotted Woodpecker.
Occurs in many places, but is nowhere common.

D. minor (Linn.). Lesser Spotted Woodpecker. „   „

Alcedo ispida, Linn. Kingfisher. Not uncommon.

Coracias garrulus, Linn. Roller. A very rare straggler. An
example obtained near Oakington in October 1835 is in
the University Museum.

Upupa epops, Linn. Hoopoe. An occasional visitant.

Cuculus canorus, Linn. Cuckoo. Very common in summer.

Strix flammea, Linn. Barn Owl. Abundant.

Asio otus (Linn.). Long-eared Owl. Local.

A. accipitrinus (Pallas). Short-eared Owl. See above.

Syrnium aluco (Linn.). Decidedly uncommon.

Athene noctua (Scopoli). Little Owl. A rare straggler, probably from the colony established by the late Lord
Lilford in Northamptonshire.

Circus aeruginosus (Linn.). Marsh-Harrier. No longer nests
in Cambridgeshire, if anywhere in England. It does not
seem to have been obtained in the county of late
years.

C. cyaneus (Linn.). Hen-Harrier. The same may be said of
this species as of the preceding.

C. cineraceus (Montagu). Montagu's Harrier. See above.

Buteo vulgaris, Leach. Common Buzzard. Now a very
rare straggler.

B. lagopus (J. F. Gmelin). Rough-legged Buzzard. „   „

Haliaëtus albicilla (Linn.). White-tailed Eagle. Occasionally
strays to the county.

Astur palumbarius (Linn.). Goshawk. A young bird, said to
have been obtained in Cambridgeshire, is in the Saffron
Walden Museum.

Accipiter nisus (Linn.). Sparrow-Hawk. No longer common.

Milvus ictinus, Savigny. Kite. Formerly common, but long
ago exterminated.

Pernis apivorus (Linn.). Honey-Buzzard. Has occurred as a
summer migrant and may still do so.

Falco peregrinus, Tunstall. Peregrine Falcon. A rare straggler.

F. subbuteo, Linn. Hobby. A summer migrant, which has bred with us.

F. aesalon, Tunstall. Merlin. A rare winter visitor.

F. tinnunculus, Linn. Kestrel. Not uncommon.

Pandion haliaëtus (Linn.). Osprey. An exceptional winter visitor.

Phalacrocorax carbo (Linn.). Cormorant. Strays now and then to the county.

P. graculus (Linn.). Shag.   „   „

Sula bassana (Linn.). Gannet.   „   „

Ardea cinerea, Linn. Common Heron. There is a heronry at Chippenham Park, and a single nest has been observed near Wisbech.

A. purpurea, Linn. Purple Heron. Used to occur in the Fen districts, but no specimens seem to have been noticed of late years.

A. alba, Linn. Great White Heron. A single example was shot on Thorney Fen in 1849 (cf. *Tr. Norfolk Soc.* v. p. 186, and *Zoologist,* 1849, p. 2568).

A. garzetta, Linn. Little Egret. A specimen from near Whittlesey, mentioned by the late Lord Lilford (*Birds of Northants.* II. p. 118), may have occurred within the county boundaries.

A. ralloïdes, Scopoli. Squacco Heron. Said to have been obtained in Cambridgeshire by Jenyns (*Manual British Vertebrate Animals,* p. 189).

Nycticorax griseus (Linn.). Has occurred near Wisbech (*Zoologist,* 1849, p. 2568).

Ardetta minuta (Linn.). Little Bittern. See above.

Botaurus stellaris (Linn.). Common Bittern. See above.

Platalea leucorodia, Linn. Spoonbill. Three specimens were killed near Wisbech in 1845 according to Jenyns (cf. *Fauna Cantabrigiensis* MS. in the University Museum of Zoology).

Anser cinereus, Meyer. Grey-lag Goose. See above.

A. albifrons (Scopoli). White-fronted Goose. A rare winter visitor.

A. segetum (J. F. Gmelin). Bean Goose. *A. brachyrhynchus*, Baillon. Pink-footed Goose. No doubt both these species occur at times in winter, but the individuals killed have not yet been sufficiently identified.

Bernicla leucopsis (Bechstein). Bernacle Goose. Said to have been obtained in the Fens and near the coast.

B. brenta (Pallas). Brent Goose. This species has occurred in the county, and may be more common than is supposed towards the Wash.

Cygnus musicus, Bechstein. Whooper. Rare, but sometimes observed in flocks in winter.

C. bewicki, Yarrell. Bewick's Swan. A rare straggler.

C. olor (J. F. Gmelin). Mute Swan. Kept on the Cam and elsewhere.

Tadorna cornuta (S. G. Gmelin). Common Sheld-Duck. An occasional winter visitor.

Anas boscas, Linn. Mallard or Wild Duck. Not very common.

A. strepera, Linn. Gadwall. Occasionally met with.

Spatula clypeata (Linn.). Shoveler. Occurs from time to time, and may possibly breed in the county.

Dafila acuta (Linn.). Pintail. A rare straggler to us in winter.

Nettion crecca (Linn.). Teal. Somewhat rare, but probably breeds locally.

Querquedula circia (Linn.). Garganey. A rare summer or possibly only autumn visitor.

Mareca penelope (Linn.). Wigeon or Whew. A winter visitor, not abundant.

Fuligula ferina (Linn.). Pochard. A winter visitor. Its breeding range may be found to extend to Cambridgeshire in the future.

F. nyroca (Güldenstadt). Ferruginous Duck. Has occurred once in the county. (See Yarrell's *British Birds*, ed. iv.

IV. p. 418 and Jenyns' *Fauna Cantabrigiensis* MS. in the University Museum of Zoology.)

F. cristata (Leach). Tufted Duck. The same may be said of this species as of the Pochard.

F. marila (Linn.). Scaup-Duck. Reported from the Fens and elsewhere in winter.

Clangula glaucion (Linn.). Golden-eye. An occasional winter visitor.

Harelda glacialis (Linn.). Long-tailed Duck. Has been shot near Ely, but there appear to be no modern records.

Oedemia nigra (Linn.). Common Scoter. A winter straggler from the coast. It is often common in the Wash in company with the Velvet Scoter or even the Surf Scoter.

O. fusca (Linn.). Velvet Scoter. Jenyns records this species from Haddenham (*Fauna Cantabrigiensis* MS.).

Mergus merganser, Linn. Goosander. A rare winter visitor.

M. serrator, Linn. Red-breasted Merganser. „ „

M. albellus, Linn. Smew. „ „

Columba palumbus, Linn. Ring-Dove or Wood-pigeon. Common.

C. oenas, Linn. Stock-Dove. Abundant.

Turtur communis, Selby. Turtle-Dove. A common summer visitor, often very plentiful.

Syrrhaptes paradoxus (Pallas). Pallas's Sand Grouse. Occurred not uncommonly in the well-known irruptions of 1863 and 1888[1].

Phasianus colchicus, Linn. Pheasant. Common, but the typical form seldom met with, nearly all individuals exhibiting the ring on the neck characteristic of *P. torquatus*.

Perdix cinerea, Latham. Partridge. Most abundant.

Caccabis rufa (Linn.). Red-legged Partridge. Abundant. It was first introduced into Suffolk in 1770, and reached Cambridgeshire by 1821, if not earlier.

Coturnix communis, Bonnaterre. Quail. Not common and seldom remaining for the winter.

---

[1] A stray example of the Red Grouse, *Lagopus scoticus* (Latham), from Histon has been reported to me by Mr W. Farren of Cambridge.

Crex pratensis, Bechstein. Land-Rail or Corn-Crake. A
common summer visitor.

Porzana maruetta (Leach). Spotted Crake. See above.

P. parva (Scopoli). Little Crake. An exceptional summer
visitor, twice recorded from the county. (See Yarrell's
British Birds, ed. iv. III. p. 149.)

P. bailloni (Vieillot). Baillon's Crake. See above.

Rallus aquaticus, Linn. Water-Rail. An occasional pair may
still breed in Cambridgeshire.

Gallinula chloropus (Linn.). Moorhen or Water-Hen. Common.

Fulica atra, Linn. Coot. Never common and breeds only
in a few places in Cambridgeshire.

Grus communis, Bechstein. Crane. See above.

Otis tarda, Linn. Great Bustard. See above.

O. tetrax, Linn. Little Bustard. A rare winter visitor.

Oedicnemus scolopax (S. G. Gmelin). Stone-Curlew. Once
abundant, but now an uncommon summer visitor, though
its nest may occasionally be found in the county.

Glareola pratincola (Linn.). Pratincole. One example has
been recorded from Cambridgeshire. (See Yarrell's *British
Birds*, ed. iv. III. p. 233.)

Eudromias morinellus (Linn.). Dotterel. Occurs on the
spring and autumn migrations, but is now rare.

Aegialitis hiaticola (Linn.). Ringed Plover. An uncommon
straggler.

Charadrius pluvialis, Linn. Golden Plover. Not uncommon
from autumn onwards, often in large flocks.

Squatarola helvetica (Linn.). Grey Plover. Occurs rarely
near the coast.

Vanellus vulgaris, Bechstein. Lapwing or Peewit. Common
in many places.

Strepsilas interpres (Linn.). Turnstone. Recorded from the
Cam during the first week of May. (*Zoologist*, 1849,
p. 2497.)

Haematopus ostralegus, Linn. Oyster catcher. A rare
straggler.

Recurvirostra avocetta, Linn. Avocet. Said by Donovan to have bred in Cambridgeshire, but this is doubtful.

Phalaropus fulicarius (Linn.). Grey Phalarope. A very rare visitor. A pair were shot some years ago in autumn between the Boathouses and Magdalene College Garden.

P. hyperboreus (Linn.). Red-necked Phalarope. Has once occurred near Cambridge in autumn.

Scolopax rusticula, Linn. Woodcock. Not common and not known to breed with us.

Gallinago major (J. F. Gmelin). Great Snipe. A specimen from "Cambridgeshire" is in the Saffron Walden Museum, and the bird has occurred in the Fens.

G. caelestis (Frenzel). Common Snipe. Plentiful.

G. gallinula, Linn. Jack Snipe. Fairly common.

Tringa alpina, Linn. Dunlin. A rare straggler in autumn and winter. It used to occur in the Fens in summer.

T. minuta, Leisler. Little Stint. First recorded in Britain from Cambridgeshire, but apparently never observed there since. (See Yarrell's *British Birds*, ed. iv. III. p. 386, and *Zoologist* 1849, p. 2623.)

T. temmincki, Leisler. Temminck's Stint. A very rare winter visitor.

T. subarquata (Güldenstadt). Curlew-Sandpiper. Has occurred on several occasions in the county, as at Guyhirn in September (*Zoologist*, 1851, p. 3279), and at the sewage farm near Cambridge in September, 1896. It has also been shot on the Cam.

T. canutus, Linn. Knot. An occasional straggler to the county, usually towards winter.

Machetes pugnax (Linn.). Ruff and Reeve. See above.

Tringites rufescens (Vieillot). Buff-breasted Sandpiper. The only Cambridgeshire specimen constituted the first record for Britain. (*Tr. Linn. Soc.* XVI. p. 109.)

Bartramia longicauda (Bechstein). Bartram's Sandpiper. Once recorded from the county. (See Yarrell's *British Birds*, ed. iv. III. p. 441.)

Totanus hypoleucus (Linn.). Common Sandpiper. A summer visitor, which may possibly breed in the county.

T. glareola (J. F. Gmelin). Wood-Sandpiper. A rare autumn visitor.

T. ochropus (Linn.). Green Sandpiper. A rare spring and autumn visitor.

T. calidris (Linn.). Common Redshank. Not uncommon, but local.

T. fuscus (Linn.). Spotted or Dusky Redshank. A rare autumn visitor, generally seen singly, but exceptionally in company.

T. canescens (J. F. Gmelin). Greenshank. An uncommon autumn visitor.

Limosa lapponica (Linn.). Bar-tailed Godwit. An autumn and spring visitor, not common.

L. belgica (J. F. Gmelin). Black-tailed Godwit. See above.

Numenius arquata (Linn.). Common Curlew. Apparently a rare visitor towards autumn; but it may be more common than is supposed towards the Wash.

N. phaeopus (Linn.). Whimbrel.   „   „

Hydrochelidon nigra (Linn.). Black Tern. See above.

Sterna fluviatilis, Naumann. Common Tern. A straggler.

S. macrura, Naumann. Arctic Tern. Recorded from Foulmire by Jenyns (*Fauna Cantabrigiensis* MS. in the University Museum of Zoology). Other species of the genus probably occur in the county, but need recording.

Xema sabinii (Joseph Sabine). Sabine's Gull. A single specimen has been recorded. (See Yarrell's *Brit. Birds*, ed. i. III. p. 423.)

Larus minutus, Pallas. Little Gull. Recorded from Wisbech by Jenyns (*Fauna Cantabrigiensis* MS. in the University Museum of Zoology).

Larus ridibundus, Linn. Black-headed Gull. Frequently observed, as a rule towards winter.

L. canus, Linn. Common Gull.   „   „

L. argentatus, J. F. Gmelin. Herring-Gull. Occurs occasionally.

L. fuscus, Linn. Lesser Black-backed Gull.   „   „

L. marinus, Linn. Great Black-backed Gull.   „   „

L. glaucus, O. Fabricius. Glaucous Gull.   „   „

Rissa tridactyla (Linn.). Kittiwake. Usually occurs in hard winters, but is never common.

Stercorarius pomatorhinus (Temminck). Pomatorhine Skua. A very rare straggler.

Alca torda, Linn. Razor-bill. Occasionally met with inland and near the coast.

Uria troile (Linn.). Common Guillemot. Strays at times to Cambridgeshire.

U. bruennichi, E. Sabine. Brünnich's Guillemot. Has been obtained once near Guyhirn (*Zoologist,* 1895, p. 109).

Mergulus alle (Linn.). Little Auk. Met with in exceptionally hard winters, sometimes far from the coast. Such winters were those of 1841—2, 1846—7, 1857, 1862—3, and 1894—5.

Fratercula arctica (Linn.). Puffin. A rare straggler.

Colymbus septentrionalis, Linn. Red-throated Diver. Various Divers are not uncommon in the Wash, most of which probably belong to this species. Supposed occurrences of the great Northern Diver and of the Black-throated Diver are given by Jenyns (*Fauna Cantabrigiensis* MS. in the University Museum of Zoology).

Podicipes cristatus (Linn.). Great Crested Grebe. Occurs but rarely.

P. griseigena (Boddaert). Red-necked Grebe. Still rarer than the preceding.

P. auritus (Linn.). Slavonian Grebe. Certainly occurs at intervals, while the Eared Grebe (*P. nigricollis,* C. L. Brehm.) may do so.

P. fluviatilis (Tunstall). Little Grebe or Dabchick. Decidedly uncommon.

Procellaria pelagica, Linn. Storm-petrel. Occasionally driven inland by severe weather to Cambridgeshire.

Occanodroma leucorrhoa (Vieillot). Leach's Fork-tailed Petrel. A specimen from Bassingbourn is in the University Museum of Zoology.

Diomedea melanophrys, Boie. Black-browed Albatross. An example was captured at Streetly Hall Farm, Linton, on July 9th, 1897 (see *Ibis,* 1897, p. 625).

REPTILIA AND AMPHIBIA OF CAMBRIDGESHIRE.

By H. Gadow, Ph.D., M.A., F.R.S., King's College.

Of the thirteen different species of Amphibia and Reptiles of England eleven are known to occur, and to breed, in Cambridgeshire. In the following pages no notice is taken of specimens which have obviously escaped from captivity. For instance, Prof. Henslow in 1824 met with a Natterjack in the Old Botanic Garden, the present Museum-grounds. In the present Botanic Garden now and then a continental frog has been found, and at the Leys School opposite boys occasionally bring such a frog from London and keep it for a while. A spotted or Fire Salamander has crawled across Chesterton Road after a thunderstorm, and inquiry into its ownership was of no avail. I mention this case because a gentleman, who knows this species from continental experience, has assured me that he has met several of them at a certain place on the East Coast.

REPTILIA.

Of the three British species the only snake of Cambridge and its neighbourhood for many miles around is the *Common* or *Grass-snake, Tropidonotus natrix*. It is easily recognisable by the pair of light-coloured, yellow or white, patches on the neck, immediately behind the head ; this light collar is always present in the snakes of this county. The Grass-snake prefers moist, grassy localities, with the neighbourhood of water, chiefly on account of the food, which consists entirely of fishes and amphibians, notably of frogs ; toads are occasionally eaten, but mice are never taken, although a Radiograph was once exhibited at a learned Society's Meeting showing a Grass-snake in the act of swallowing a mouse, but *horribile dictu,* the mouse had been shoved by force into the unwilling snake's mouth! The Grass-snake never bites, although hissing and striking out furiously with its head. Its only defence, when caught, consists of the contents of the cloaca and the anal glands, the voided mess having an abominable odour.

Pairing takes place here in May or June; the numerous eggs are laid from July to the end of August, in mould, heaps of weeds, or in manure heaps. The eggs are not laid in strings, but they soon stick together by contact. As a rule they do not contain any visible sign of the embyro, but it often happens that the snake has to delay oviposition, as often occurs with captive specimens, and in such a case the embryos are more or less advanced.

Grass-snakes used to occur not unfrequently along the meadows of the Cam, for instance at Coe-fen, and even at the Backs of the Colleges. In the Fens they are still common.

Coronella laevis, the Smooth snake, does not occur in East Anglia.

Vipera verus, the *Viper,* ought to be a very common species in this county, considering the many localities which would at first sight seem to be well suitable for this creature. But it is a rare species. I myself have never found one, and I do not know of a single instance of the occurrence of a specimen in the county, either on the sunny slopes of the chalky Gog-Magogs, or on the Boulder Clay towards Essex, or in the clammy Fens. The Rev. L. Jenyns wrote in 1835 "In Cambridgeshire very rare," and in 1859 "Has occurred in a few instances in the neighbourhood of Cambridge, but is apparently very rare in the county, and I have never met with it myself."

Lacerta vivipara, the *Common English Lizard*, is, according to the Rev. L. Jenyns, "very common on banks, heaths, and other open places; often seen in the drier parts of the Fens...the young broods appearing in June and July." It is, however, very local, for instance near Fulbourn, Wilbraham Fen and at Sawston Moor. I do not know of an instance of its occurring on the Gog-Magogs, at Grantchester, or at Madingley.

Lacerta agilis, the *Sand Lizard*. Whilst the Common Lizard occurs throughout Great Britain and even in Ireland, the Sand Lizard seems to be restricted to the southern half of England, where it prefers sandy heaths, the edges of copses, railway-banks and similar, rather dry, localities. Its occurrence near Cambridge, e.g. along the Devil's Ditch, near Newmarket, and thence into Norfolk, seems to mark the northern limit of this species. But there is no doubt that *L. agilis* is often confounded with *L. vivipara*. Jenyns, for instance, did not distinguish between them.

The main differences are the following: *L. vivipara* is viviparous; has a single postnasal and a single anterior loreal shield; and the supra-ocular and supraciliary scales are in contact with each other; the coloration, although subject to much variation, is brown to reddish above, with small darker and lighter spots; often with a blackish vertebral streak and a dark lateral band edged with yellow; underparts orange to red in the male with conspicuous black spots; yellow or pale orange in the female, with or without scanty black spots.

L. agilis lays eggs; has usually a single postnasal but two superposed anterior loreals, the three forming a triangle. The supraoculars are separated from the supraciliary scales by a series of little granules. As a rule the Sand Lizard appears longitudinally striped, owing to rows of dark and white spots along the sides of the back, flanks and tail. The prevailing ground-colours are a more or less pronounced green in the male (hence often called "Green Lizard"), in the female brown and grey. Underparts in males yellow with black spots, in females cream-coloured.

Anguis fragilis, the Blindworm, or Slow-worm, is very local and rare in this county. Jenyns "only in a few instances noticed it about Bottisham."

AMPHIBIA.

All the three British Newts occur in Cambridgeshire, and during the pairing season all three may be found in the same pools; for instance, between Chesterton and Milton, on grasslands where small pools have been dug for the cattle, down to the permanent water-level, or on the higher ground, where the impervious gault secures the presence of water. Not a few of these little ponds are surrounded with trees by the roots of which their over-hanging banks are prevented from tumbling down, and here the water is teeming with insect life. In such places Newts not only hibernate regularly, especially the Crested Newt, but the young, although attaining to a considerable size, partly retard their metamorphosis, in so far as they retain portions of their external gills in the shape of short, scarcely serviceable fringes, far into the autumn, but by the next spring every trace of the clefts and gills is lost.

Triton vulgaris (s. *taeniatus*, s. *punctatus*), the *Common*, or *Spotted*, or *Smooth Newt* is the most frequent, often occurring in ditches, for instance, near Grantchester. Its usual length does not surpass 3 inches, and it is easily distinguished by the yellow to orange undersurface, which is always spotted with black. The nuptial dress of the male shows a non-serrated, high and very wavy crest which extends from the neck without interruption into the tail-fin which is also wavy.

Triton cristatus, the *Crested* or *Warty Newt*, when full adult averages 5 to 6 inches. The skin is slightly tubercular, and there is a strong gular fold. The underparts are pale yellow, almost always with large, black spots. The breeding male is very beautiful. A high, serrated or jagged crest on

the head and trunk, separated by a gap from the entire crest of
the tail, the sides of which are adorned with a bluish white
band. The female has a yellow line along the middle of the
back.

Triton palmatus (s. *helveticus*), the *Webbed Newt*, so
called because the toes are fully webbed. The smallest of
European newts, remaining mostly under 3 inches in length.
The tail ends in a thread, which in some males is almost
half-an-inch long, while it is only just indicated in the female.
A similar tail filament exists in the two other newts during
the larval stage. The male of the webbed newt develops
during the breeding season a cutaneous fold along each side
of the back, but only a low, entire vertebral crest. This little
newt is rather common in Quy Fen.

It may be added that the skin of all newts, especially that
of *T. cristatus*, contains numerous poison glands. If anyone
doubts the efficiency of the secretion he can easily, with
great discomfort to himself, be convinced of the fact by
rubbing a living newt between the fingers and then applying
his tongue to them.

Bufo vulgaris, the *Common Toad*, is practically ubiquitous,
but remains rather small, seldom attaining the length of
3 inches. Those specimens which live on chalk soil assume
a pale olive grey, a much lighter colour than those which
inhabit other kinds of terrain, where brown or rusty tints
prevail.

Bufo calamita, the *Natterjack*, or *Running Toad*, is very
local, being restricted to sandy localities. It was first found
in considerable abundance on Gamlingay Heath, in 1824, by
Henslow and Jenyns. Gamlingay is still the chief locality
in the county where this little toad is frequent. During the
months of May and June they there resort to certain clay-pits,
in the shallow water of which they spawn between the rushes.
They hibernate in deep holes, either dug by themselves, or
appropriating those of the sand-martins, which have established
a colony in the steep and high walls of the sand which there

overlays the clay. Another place of occurrence is Tadlow on the Bedford Road.

Rana temporaria, the *Common* or *Grass-frog*, is found throughout the county. In the vicinity of Cambridge it is becoming decidedly rarer owing to the incessant demand for this martyr of science. Within the last twenty years the price has risen from one half-penny to a penny per piece, and the purveyors have to extend their raids further and further a-field.

Rana esculenta, the *Edible* or *Water-frog*. The Fens of Cambridgeshire and Norfolk seem to be the only districts in England which may rightly claim this otherwise continental species as indigenous. Bell (*British Reptiles*, 2nd ed. 1894, p. 110) records that his father, who was a native of Cambridgeshire, had noticed the presence of these frogs, many years before the publication of the book cited, at Whaddon and Foulmire : they were known from their loud croak as "Whaddon organs" and "Dutch nightingales." But the species was not officially discovered until September, 1843, when Mr Charles Thurnall, of Duxford, caught and sent two specimens to the British Museum. In the following summer more were caught in the same Foulmire Mere, and some of these are preserved in the Museum of Zoology at Cambridge. Next, the species was rediscovered in Norfolk, between Thetford and Scoulton, where it is very abundant, and from inquiries made by Lord Walsingham must have existed at least since about 1820. Occasionally a specimen is caught in some other locality; for instance, in the Fens of Foulden, Norfolk; I have heard them in the pairing season of 1883 on Hickling Broad; and in the summer of 1901 one fine male was found amongst a number of Grass-frogs which had been caught between Chesterton and Milton for the Physiological Laboratory.

The interesting question is, of course, whether these Water-frogs have been introduced by man, or whether they are the last lingering descendants of indigenous Britishers, survivors from a time when *R. esculenta* was a normal inhabitant of this country. Fossil or semi-fossil remains from the Fens are unknown. Great numbers of Water-frogs, imported from Belgium and France, were turned loose in the Fens of Foulden and other places in the years 1837, 1841, and 1842[1]. These Foulden specimens, or at least their direct descendants, belong to the *var. typica* of *R. esculenta*. Those, however, which have been found in Foulmire, Thetford, Scoulton, and Chesterton belong undoubtedly to the *var. lessonae*, which is distinguished by the much stronger inner metatarsal tubercle, while the outer one is almost vanishing, by the proportionally larger fourth toe, and by the general coloration.

Now begins the muddle. When Mr Boulenger ("Notes on the Edible Frog in England," *P.Z.S.* 1887, p. 573) had discovered that the Foulmire frogs belonged to the *var. lessonae*, which was then believed to be confined to Italy, he naturally suggested that the English specimens were of Italian origin, perhaps introduced by the monks. But it is now known that the *var. lessonae* has a sporadic and much wider distribution, it having been found not only in Austria, Hungary, Bavaria and various other parts of Germany, but also, by Boulenger himself, in Belgium and near Paris. The supposed Italian origin of our frogs has naturally lost its interest by these recent discoveries, but, nevertheless, we must remember that there existed considerable intercourse between East Anglia and the monks of Lombardy, who, to mention only one instance, came regularly to the old Priory of Chesterton, in order to collect their rents. If the frogs were introduced by them for culinary purposes into various suitable localities, their descendants would remain as local as they actually are and as are also the undoubtedly introduced French specimens of the *var. typica*. On the other hand, if we assume the *lessonae* specimens to be the last living descendants of English natives, it is

---

[1] Great numbers have within the last few years been introduced from Germany, Belgium and Italy, and have been liberated in various counties, e.g. Surrey, Hampshire, Oxfordshire and Bedfordshire.

inconceivable why they should now be restricted to that eastern corner, while there are hundreds of other suitable places, which, if on the Continent, would be perfect paradises for Water-frogs.

However, we know next to nothing about the oecology of these creatures. It is quite possible that the sporadic occurrence of the *var. lessonae* is due to local adaptation and changes of the typical form, wherever the same favorable, or necessary, but to us unknown, conditions prevail. And after all, the differences between these two varieties are not great, and in many specimens are even arbitrary, just as we might expect in actually changing forms. On the other hand, we know of a good many species which are either actually spreading, or which—and this may apply to the present case—are slowly but surely vanishing. In prehistoric times *Emys europaea*, the *Pond-tortoise*, was common in the Fens. It has now receded eastwards, being extinct in Belgium, Holland and west of the Elbe; between this river and the Oder it is now in the vanishing stages, while it is still plentiful in Poland and Russia. *Tempora mutantur, nos et mutamur in illis.*

THE FISHES OF CAMBRIDGESHIRE.

By E. VALLÉ-POPE, M.A., Emmanuel College.

CAMBRIDGESHIRE is eminently a county of anglers. This fisherman instinct comes from bygone generations, for fishing was once an industry of the county and adjoining districts. It was one of the objections raised against the first drainage of the Fens that many thousands engaged in fishing and fowling would be thrown out of employment. As late as 1749 there was a fish market in Cambridge itself, which was supplied with fresh-water fish from the neighbouring Fens. It was held twice a week, and salmon and sturgeon could be purchased.

In early days the Cambridgeshire fisheries were numerous and important, a considerable portion of the endowments of the old monastery of Ely being derived from them. From accounts in Dugdale and Camden the amount of fish was enormous. According to Bede the very name of Ely itself is derived from the vast number of eels caught in the vicinity. The importance of these fisheries in ancient days was so great that lawsuits were waged over them. On one occasion the Abbot of St Edmondsbury successfully obtained an injunction against the diverting of the Nene requisite to protect Wisbech and the adjoining country from inundation. The Abbot pleaded that should the course of the Nene be altered, a certain fishery in that district belonging to his Abbey would be ruined. With reference to later times a quotation from an old fragmentary History of Cambridgeshire may be instructive. It was written in 1749 by one Edmund Carter, and contains a description of the old Cambridge Fish Market of some antiquarian interest. It reads as follows : "The Fish

Market which is separated from the Herb Market by Butcher Row is on Wednesdays and Fridays sufficiently stocked with Fish from the neighbouring Fenns and sea fish from Lynn. Fresh Salmon and Sturgeon is sometimes also brought to the Market and usually sold at about twelve pence the pound......
.........but on the chief market day there is seldom any fish to be had but Eels and Jacks, which are extraordinary cheap." Such was the Cambridge fish market in the days when the navigation of the Cam was frequently stopped for some time with the mass of boats and barges conveying merchandise to the town, showing that there was still reason for the arms of the town—three ships with Neptune's horses as supporters.

Cambridgeshire and the Fen district is essentially the home of the coarse fish. A line drawn from the Humber to the Dorset coast roughly marks out their habitat and separates them from their more stately brethren of Northern and Western England—the Salmonidae. No portion of this triangular space contains such an abundance of fish as the Fenland. A catch of bream in the Norfolk Broads is reckoned by the stone. A take of eight or ten pike in one day by a single rod is no rare occurrence. That fish should abound in East Anglia is but natural. The vast system of drains, cuts, and waterways, so numerous that the natural course of the rivers has been almost obliterated, affords a home and shelter for fish unequalled in England. On the other hand, it must be remembered that the drainage system of the Fens has been largely responsible, in all probability, for drying up many of the smaller drains and channels and leaving them completely without water during the summer. Consequently there has been a certain depletion of fish. At least, such is Mr Miller's theory in *Fenland*, and I am disposed to agree with him, though I cannot accept as an important destructive agent the admission of salt water into the larger drains, which Mr Miller mentions as a cause of injury to the fresh-water fish. On the whole the fishes of Fenland enjoy comparative immunity from harm. There are no large manufacturing towns on the banks of our East

Anglian rivers to poison the water with pollutions, and along
the waterways there is no extensive traffic.   The two enemies
are the angler and the otter.   With regard to the fish themselves their history has been very uniform.   The drainage of
the great meres of Ramsey and Whittlesea has been responsible
for the disappearance of many birds and insects—the large
Copper (*Polyommatus dispar*) amongst them.   But the fish
have remained unaltered according to all records, unless the
Barbel (*Barbus vulgaris*) ever existed in the waters of the
old Fens.   It certainly does not exist now so far as I am
aware.   An abortive attempt was once made to introduce
them into the Great Ouse, but they were never heard of or
seen again.

There is extant to my knowledge no list of fishes belonging
to the county.   There is, however, a list of Fenland fishes in
Messrs Miller and Skertchly's *Fenland,* to which I am much
indebted, and I derived much assistance from the collection of
fish in the Museum at Wisbech.   In all there are, including
the Lamprey and reckoning the denizens of fresh as well as
salt water, forty-two species of Cambridgeshire fishes.   When-
ever a mention is made in the list of an isolated instance,
the reference is to some specimen preserved in the Wisbech
Museum.

TELEOSTOMI.

Acanthopterygii.

1. Perch, *Perca fluviatilis* (Linn.).   Large perch are taken
    in the Great Ouse.
2. Ruffe, *Acerina cernua* (Linn.).   An interesting Cambridge
    fish, as according to a story (I believe erroneous) it was
    discovered by Dr John Caius.
3. Sea Bream, *Pagellus centrodontus* (Cuv. et Valence).
    Doubtful.   One specimen taken in Pauper's Cut.
4. Horse Mackerel, *Caranx trachurus* (Cuv. et Valence).
    One specimen taken in the Wisbech River, 1854.

5. Dory, *Zeus faber* (Linn.). Two specimens have been taken, one in the Wisbech River, 1884, and one at Wisbech Eye, 1854.
6. Great Weaver, *Trachinus draco* (Linn.). One specimen taken in the Wisbech River, 1884.
7. Angler, *Lophius piscatorius* (Linn.). One specimen caught in the Nene.
8. Miller's Thumb, Tommy Logge or Bullhead, *Cottus gobio* (Linn.).
9. Red Gurnard, *Trigla cuculus* (Linn.). One specimen taken in the Wisbech River.
10. Viviparous Blenny, *Zoarces viviparus* (Cuv.). Doubtful. There is a specimen in the Wisbech Museum, captured 1842, but no locality is mentioned.
11. Gar Fish, *Belone vulgaris* (Cuv.). One taken at Wisbech.
12. Gemmeous Dragonet, *Callionymus lyra* (Linn.). One taken below Wisbech town, 1855.
13. Three-spined Stickleback, *Gasterosteus aculeatus* (Linn.).
14. Ten-spined Stickleback, *Gasterosteus pungitius* (Linn.). A friend of mine captured a specimen at Upware.
15. Lesser Grey Mullet, *Mugil chelo* (Cuv.). Sometimes occurs at Wisbech.

ANACANTHINI.

16. Burbot or Eel Pout, *Lota vulgaris* (Cuv.). There is one in the Wisbech Museum which weighed 5½ lbs.
17. Flounder, *Pleuronectes flesus* (Linn.).

PHYSOSTOMI.

18. Carp, *Cyprinus carpio* (Linn.).
19. Crucian Carp, *Cyprinus carassius* (Linn.). The Gold fish (var. *Auratus*) occurs.
20. Gudgeon, *Gobio fluviatilis* (Günther).
21. Roach, *Leuciscus rutilus* (Linn.).
22. Chub, *Leuciscus cephalus* (Linn.).
23. Dace, *Leuciscus dobula* (Linn.).
24. Minnow, *Leuciscus phoxinus* (Linn.).

25.　Rudd, *Leuciscus erythrophthalmus* (Linn.).
26.　Tench, *Tinca vulgaris* (Cuv.).
27.　Bream, *Abramis brama* (Linn.).
28.　White Bream or Breamflat, *Abramis blicca* (Cuv.). This
　　　species was first observed in this country in 1824.
　　　Specimens were obtained from the Trent. The Rev.
　　　L. Jenyns subsequently found the fish existing in the
　　　Cam. It is a much smaller species than the common
　　　Bream and far less widely distributed. Mr Brindley of
　　　St John's College once took a specimen of the White
　　　Bream normal in every way save that it lacked the
　　　pelvic fins and girdle: it was nevertheless in as good
　　　condition as others obtained from the same shoal.
　　　This abnormal fish has been placed in the Museum
　　　of the Royal College of Surgeons.
29.　Pomeranian Bream. Stated to exist in the county by
　　　Houghton and Cholmondley Pennell.
30.　Bleak, *Alburnus lucidus* (Günther).
31.　Loach, *Nemachilus barbatulus* (Günther).
32.　Spined Loach, *Cobitis taenia* (Linn.). Yarrell says it
　　　has been found in Cambridgeshire.
33.　Pike, *Esox lucius* (Linn.).
34.　Allice Shad, *Clupea alosa* (Linn.). Doubtful. There is
　　　a specimen in the Wisbech Museum, but the locality of
　　　capture is not mentioned.
35.　Twaite Shad, *Clupea finta* (Linn.). Doubtful as a Cam-
　　　bridgeshire fish. One specimen was taken, 1854, in
　　　the Nordelph River.
36.　Salmon Trout, *Salmo trutta* (Linn.). One specimen,
　　　1849, Wisbech.
37.　Trout, *Salmo fario* (Linn.). Plentiful in south of county.
38.　Smelt, *Osmerus eperlanus* (Linn.).
39.　Eel, *Anguilla vulgaris* (Turt.).

CHONDROSTEI.

40.　Sturgeon, *Acipenser sturio* (Linn.). Occasional. I have
　　　heard of seven specimens being taken in the Great

Ouse within living memory. Two have been seen at
St Ives this year. In early times there are records of
" greater and royal fish " existing at Ely.

ELASMOBRANCHII.

41. Angel or Monk Fish, *Squalus squatina* (Linn.). One
    was taken near Wisbech, 1882.

CYCLOSTOMATA.

42. Sea Lamprey, *Petromyzon marinus* (Linn.). There is a
    specimen in the Wisbech Museum, caught in the Nene.
    [I have been informed that the Lampern, *Petromyzon
    fluviatilis*, used to be found in the Cam. I can find
    no authority however to warrant my placing it on the
    list as existing now. Dr W. Gaskell knew an old
    inhabitant of Shelford who told him that the lamprey
    was in his earlier days plentiful at that place. It is of
    course still abundant in the Little Ouse, at Brandon
    Creek, and also between Houghton Lock and
    Huntingdon.]
    The above are fishes which may be considered as belonging
to Cambridgeshire. I append below a small list of fishes which
are stated to have been taken in the Fenland.
    Grey Mullet, *Mugil capito* (Cuv.). On the authority of
    Mr Miller.
    River Lamprey or Lampern, *Petromyzon fluviatilis*
    (Linn.).
    Pipe fish, *Syngnathus acus*. One taken at Sutton Bridge,
    1836.
    Sea Scorpion, *Cottus scorpius*. Sutton Wash.
    Gorgeous Opah, or King fish, *Lampris luna*. A specimen
    of this bizarre looking fish was taken at Hunstanton
    in 1839, and is preserved in the Wisbech Museum.

THE MOLLUSCA OF CAMBRIDGESHIRE.

By H. H. Brindley, M.A., St John's College.

The Mollusca are fairly well represented in the county of Cambridge, since 101 of the 143 or so British species have been recorded. This is, however, a larger total than those of all the bordering counties except Essex, yet it seems probable that further search may add to the present list. Exploration has no doubt been restricted by the facts that the large towns are at the north and south ends of the county and that the intermediate region is one of small villages and scattered farmsteads. At the same time the comparatively recent alteration in the physical features of much of the county demands attention in considering its fauna. The draining of the Fenland has during the last 300 years converted vast areas of brackish tidal waters into dry land, and so the absence of at least some of the xerophilous species common enough in other parts of England is not difficult to understand.

Beyond this it would be hazardous to say much, for our knowledge of the molluscan fauna of the central districts is at present very inadequate, and much remains to be done in the exploration of the region through which the Bedford Levels run, the tidal waters of those artificial streams, and of the region forming the Isle of Ely, before we can consider that the mollusca of the county have been fully recorded. The immediate neighbourhood of Cambridge has been the scene of closer search than the Fenland to the north, and for the former our information is perhaps fairly full, though it is possible that one or more forms found in the Pleistocene gravels of the district which still live in other parts of England may prove to have survived in the county.

Nevertheless there is much to reward the searcher in the neighbourhood of Cambridge. In the Cam as it passes through the Backs and immediately below the town many of the commoner freshwater forms are to be met with, though it appears probable that the increasing pollution of the water has driven some species down stream, in the same way that the lamprey has been ousted from the river at and above Cambridge in recent years. The departure of the Unionidae is noticeable, and for them and also for *Dreissensia* the neighbourhood of Upware is a more favourable locality than Cambridge. *Anodonta cygnea*[1] occurs abundantly in the Cam above Grantchester, and *Sphaerium rivicola* may be found in many places. *Pisidium fontinale* lives in the backwaters of the Cam in the Backs and in the river itself lower down, while *P. amnicum* occurs more frequently near Upware. Among weeds in the backwaters of the river as it passes behind the Colleges there will be found in fair abundance *Bithynia tentaculata, Valvata piscinalis, Limnaea pereger, Planorbis corneus, P. albus, P. carinatus, P. umbilicatus,* which is very common, and *P. vortex. Limnaea auricularia, L. palustris* and *Segmentina nitida* occur more sparingly. *Ancylus fluviatilis* and *Acroloxus lacustris* are also to be met with without much search. *Neritina fluviatilis* is fairly numerous on the bases of the College bridges. *Physa fontinalis* is not uncommon, and *Succinea elegans* may be found anywhere among grass and wood on the banks.

---

[1] The nomenclature throughout this article is that employed by Mr B. B. Woodward in his "List of British Non-Marine Mollusca" (*Journal of Conchology*, x. p. 353, October, 1903), and adopted by the Conchological Society in their revised list.

Limnaea stagnalis is more frequently met with in waters below Cambridge, as at Burwell, Wicken and Ely. *Limnaea truncatula* seems rather local in the Cam, but it is not uncommon in ditches at Cherryhinton and in the Ouse at Ely. *Vivipara contecta* also does not seem common in the river at Cambridge, but it may be found abundantly in weedy backwaters near Clayhithe, and from there down to Ely. Bourne Brook, which runs from Caxton to the Cam at Grantchester, is also a good locality for this species. *V. vivipara* is apparently not an inhabitant of the Cam or is rare, though it occurs within the county.

Turning from the river itself we find much of interest in the molluscs to be found in the numerous ponds and ditches close to Cambridge. Just above the town are the pasture lands known as Sheep's Green and Coe Fen, and through these the Cam now flows in a course which is partly artificial. Hence many of the ponds, especially of Sheep's Green, are remains of the old bed of the river. In a paper on local Pleistocene species Mrs McKenny Hughes[1] has pointed out that in these ponds occur several of the less common species and that it is a striking fact, as shown by an investigation by Mr J. R. Le B. Tomlin, that the various patches of water exhibit great differences in their molluscan fauna, while some of them also contain forms which are either rare or else apparently absent from the river in its present condition. Thus in these pools there occur *Valvata cristata*, *Bithynia leachii*, *Planorbis crista*, *P. fontanus* and *Aplecta hypnorum*. One specimen of *Planorbis glaber* has also been found in a Sheep's Green pool. Of the forms which are found in the river, *Velletia lacustris*, *Planorbis albus* and *Segmentina nitida* also inhabit Sheep's Green. *Carychium minimum* is numerous on the grass at the margin of the pools. Mrs Hughes points out that the conditions above described

[1] "On the Mollusca of the Pleistocene Gravels in the neighbourhood of Cambridge," *Geological Magazine*, Dec. 3, v., 1888, p. 193.

throw some light on the way in which discontinuous distribution of species within a small area may arise where a river runs through it with an unrestrained and changeful course.

The ditches and ponds round Cambridge contain a fair number of species. In many of them *Planorbis albus* occurs more frequently than in the river and though abundant is rather local. *Ancylus fluviatilis* occurs in Hobson's Brook. *Limnaea peregra* is also very common, and the extreme variability of this form is well illustrated by a comparison of specimens from the local waters. Thus in a collection made in the immediate neighbourhood of Cambridge by Mr W. Bateson, F.R.S., the present writer, on a comparatively superficial survey, obtained the general results shown on pp. 118 and 119.

Place of collection	General size	Habit of shell	General colour	Mouth
Stream through Sheep's Green near the Mill	large	the heaviest	yellowish-brown, dull	typical
Brook through the middle of Coe Fen	medium	fairly heavy	brownish orange	typical
Ditch in Madingley Road near the Stone House	rather small to medium	rather thin, with white incrustation	pale horn, glossy	typical
Ditch alongside Grantchester Footpath	medium	fairly heavy	horn yellow, upper part dark and encrusted	the narrowest
Skating Pond in Grantchester Meadows	the largest	thin in relation to size	pale horn, glossy	pillar lip much reflected
Brook leaving Barton Road beyond Brick Pit, 100 yards from road	rather small	thin	horn, glossy	typical
Stream across Huntingdon Road, $\frac{1}{2}$ mile beyond Girton College	fairly large	thin	pale horn, rather dull	typical
Brook at Little Shelford	small	thin	pale	narrow

Lowest whorl	Suture	Spire	Striae	Vertical ridges	Spiral ridges
typical	rather shallow	short	well marked	well marked	variable, some having strong ridges and others of same size none
typical	shallow	rather short	distinct	fairly well marked	none
typical	shallow	rather variable, usually short	fairly well marked	none	none
the upper whorls more convex in these than in the others	deep	well marked	well marked	well marked	fairly prominent in the larger specimens
angular on pillar side	deep	well marked	deep	well marked	slight on most, prominent in large specimens
typical	deep	rather short	faint	a few on some	none
typical	shallow	produced	well marked	well marked	fairly prominent in larger specimens
typical	deep	produced, shell axis elongated	faint to fine	none	none

The wide range of variation here seen is of course not without parallel in the case of many other animals and plants, but the facts are given in some detail as they show the ease with which useful information may be obtained from common forms within a restricted area, and an instance like this furnishes a good starting-point for a systematic enquiry by anyone to whom variation is of interest. Dealt with more fully and the enquiry extended to others of the small water-courses of the neighbourhood, with regard held also to their individual features and past and present connections, a study of the local variation of this *Limnaea* might well yield an instructive result.

Mr B. B. Woodward has pointed out[1] that the Pleistocene gravels of Barnwell contain many apparent intermediates between *Limnaea peregra* and *L. auricularia,* while many of the former are the inflated form named variety *ovata* by older writers.

Passing to terrestrial species it will be found that the Backs are a fairly good hunting-ground, and here among moss and grass roots in the damper spots several of the *Vitreas* are common, such as *V. alliaria, V. crystallina, V. nitidula,* and with them *Zonitoides nitidus. Vitrea cellaria* occurs more frequently under stones. Of the Helicidae *Hygromia hispida* (among grass roots in the damp) and *Vallonia pulchella* may be found fairly abundantly.

Sheep's Green and Coe Fen also have several of the above species, *Vitrea nitida,* for instance, occurring numerously near water. *Agriolimax laevis* is also an inhabitant of these pastures.

The Botanic Garden is a fairly good locality for the commoner forms, and the writer is indebted to Mrs Hughes for the record of *Testacella haliotidea* from this part of Cambridge. It was probably introduced from Kew about twenty years ago. *Vitrea lucida* is also found in the Garden

---

[1] "Notes on the Pleistocene Land and Freshwater Mollusca from the Barnwell Gravels," *Proc. Geol. Assoc.,* x., 1888, p. 355.

and seems to have been imported, though it is perhaps indigenous in the county. White examples of *Pyramidula rotundata*, a common form in the neighbourhood, are found in the Garden. As is usual in such a locality, certain foreign species, introduced in packets of plants or seeds, have found a home in our Garden, the cases so far observed being two of the Stenogyridae, *Opeas goodalli* and *O. urichi* from the West Indies, and *Physa acuta* from the Continent.

In gardens in Cambridge and in the immediate neighbourhood many of the commoner species are easily met with. *Limax maximus* has been recorded in cellars and hothouses within the town, while *L. flavus* is frequent in cellars and damp lands. *Milax sowerbyi* and *M. gagates* have been found in one or two gardens, and the latter also by the railway bridge under Hills Road. *Arion hortensis* is also abundant. In gardens on the Huntingdon Road side of the town *Arion ater* and *Helix aspersa* are destructively common, and after showers gravel walks and the side path up to Howes Close may be covered with *Hygromia rufescens*. *Helicella caperata* is frequent among scattered straw, and *H. cantiana* is quite common in the neighbourhood. *H. itala* is more local near Cambridge, but it occurs in places, though apparently not very numerously. *Helix nemoralis* is a common form in hedgerows, *H. hortensis* occurring more locally but in fair abundance in similar situations. The former of these two is inclined to follow the course of the river. *Ena obscura* is also fairly common.

Madingley Wood is a good hunting ground and therein have been obtained *Vitrea excavata*, *Punctum pygmaeum*, *Acanthinula aculeata* (the only record for the county and first found by the Rev. A. H. Cooke in 1876) and *Clausilia laminata*. *Caecilioides acicula* has been collected by Mr Cooke from fields near Coton to the west of Madingley and also from an Anglo-Saxon burial ground on the site of Girton College.

Passing to the south-west of the above places, the woods in the neighbourhood of Grantchester are a locality for

Vallonia pulchella and *V. costata*, for which records also the
writer is indebted to the Rev. A. H. Cooke. Damp spots in
this neighbourhood contain *Succinea putris*. In the woods
round Grantchester *Helicigona arbustorum* is common, as well
as along the course of the Cam above and below Cambridge.
This species is fond of the willows and it is not unusual
to find *Helix nemoralis* at the same time. During the Long
Vacation of 1892 on a day when the river was in flood the
writer found the raised foot-path across the meadows from
Waterbeach Station to Bottisham Lock crowded with individuals of these two species, which had found refuge thereon
from the water covering the grassland on either side. Of 639
specimens of *Helicigona arbustorum* picked up haphazard from
80 yards of the foot-path 465 were adults, and of these 15
were without any trace of the band which is typical of this
species. In practically all the rest the band was well-marked
for several whorls, so that the variation was markedly discontinuous. Of 261 adult examples of *Helix nemoralis* taken
from the same spot 21 were without bands, and were mostly
brownish in colour, a few being orange or yellow. The
banded specimens forming the majority were of most of the
usual colours, while the banding varied so greatly as to defy
any summary expression, as is so often the case in this species.

In the contiguous districts of Harston, Hauxton and
Hoffer's Bridge to the south of Cambridge several of the
less common species have been found, for instance, *Vitrea
rogersi, V. radiatula, Punctum pygmaeum, Caecilioides acicula,
Vertigo antivertigo* and *V. pygmaea.* *Arion intermedius*
occurs at Whittlesford to the east of the above places.

Passing eastward from the lower lands to the Chalk
of the Gogmagog Hills and their approaches we find some of
the Helicidae very numerous. *Helicella virgata* is plentiful
in both the type forms and several of the more or less recognised varieties. *H. itala* and *H. cantiana* are also common,
while *Ena obscura* and *Cochlicopa lubrica* are not rare.
Several of the less common species have been found in the

woods which border the Roman Road over the Gogmagogs, e.g., *Sphyradium edentulum, Jaminia muscorum* and *Vertigo pygmaea.*

In the above an attempt has been made to indicate the characters of the molluscan fauna of Cambridge itself and of places within easy reach in an afternoon's excursion. Certain cases of isolated occurrence in the neighbourhood will be mentioned later on.

The molluscan fauna of Ely seems to be much the same as that of the Cambridge district. The less contaminated state of the river at the former place is probably the reason for the greater abundance of the larger bivalves there noticeable. *Vivipara contecta* and *Limnaea stagnalis* seem to occur more frequently at Ely than at Cambridge. The trees of the Cathedral Close have been mentioned by Mr L. E. Adams[1] as a good locality for the species of *Helicella*. Littleport on the Ouse five miles north of Ely is recorded as a locality for *Planorbis crista*, while *Dreissensia polymorpha* attached to living *Anodonta cygnaea* has also been obtained there.

As has already been stated the central and northern portions of Cambridgeshire seem to have been but slightly searched. Of Streptoneurous forms *Paludestrina jenkinsi* has been found in the Wisbech Canal and in the Nene near Foul Anchor Ferry, and in the last-named locality *P. stagnalis* also occurs. The Wisbech Museum has a specimen of *Vivipara vivipara* from the district. Besides most of the species already mentioned as occurring commonly or fairly commonly in the neighbourhood of Cambridge, and in the Cam and Ouse down to Ely, the Samuel Smith and other cabinets in the Museum have local examples of *Balea perversa, Clausilia bidentata, Sphaerium lacustre, Pisidium amnicum* and *P. pusillum*. *Planorbis crista* and *P. spirorbis* also occur in this district, while very fine examples of *Aplecta hypnorum* have been obtained at Grip, near March. At present, however, our knowledge of the mollusca fauna of the fen district of

[1] *Journ. Conch.,* VI., 1890, p. 277.

Cambridgeshire is too scanty to risk any comparison in detail between it and that of the higher land round and to the south of the county town.

In the above mention of the forms likely to be met with in different directions varieties have been neglected. An attempt to deal with these would prolong this account unduly, and also the very unequal values attaching to the varieties which have received names more or less recognised render it inadvisable to lay any stress upon them without at the same time weighing individual claims. Hence, in what has already been said, and in the remainder of this account, varieties of species are referred to only in a few necessary cases.

The following is a list of species which have been obtained in the living state in the county, the figures placed after the names indicating occurrence in the local Pleistocene deposits, as follows:—(1) Marine-shell-bearing gravels of the March district; (2) Gravels of the ancient river system of the Cambridge district; (3) Gravels of the present river system as exposed at Barrington, Barnwell, Grantchester, &c.; (4) Fen silt; (5) Shell marl, as at Burwell, Stretham, Soham, Lakenheath, and between Littleport and Downham in Norfolk.

GASTEROPODA.

STREPTONEURA
   ASPIDOBRANCHIATA
      RHIPIDOGLOSSA
         NERITIDAE
            Neritina (*Theodoxia*) *fluviatilis* (Linn.)
   PECTINIBRANCHIATA
      TAENIOGLOSSA
         POMATIIDAE
            Pomatias elegans (Müll.) (3)

PALUDESTRINIDAE
Bithynia tentaculata (Linn.) (1) (3) (5)
Bithynia leachii (Shepp.)
Paludestrina jenkinsi (Smith)
Paludestrina stagnalis (Baster)
VIVIPARIDAE
Vivipara vivipara (Linn.)
Vivipara contecta (Millet) (5)
VALVATIDAE
Valvata piscinalis (Müll.) (1) (3) (5)
Valvata cristata, Müll. (3) (5)

EUTHYNEURA
  PULMONATA
    BASOMMATOPHORA
      AURICULIDAE
      Carychium minimum, Müll. (3)
    LIMNAEIDAE
      Ancylus fluviatilis, Müll. (3)
      Acroloxus lacustris (Linn.)
      Limnaea palustris (Müll.) (3) (5)
      Limnaea truncatula (Müll.) (3)
      Limnaea stagnalis (Linn.) (3) (5)
      Limnaea (Radix) auricularia (Linn.) (3) (5)
      Limnaea (Radix) peregra (Müll.) (3) (5)
      Amphipeplea glutinosa (Müll.)
      Planorbis carinatus Müll. (3) (5)
      Planorbis umbilicatus Müll. [=*complanatus* Jeffreys
        =*marginatus* Draparnaud] (3) (5)
      Planorbis vortex (Linn.) (3) (5)
      Planorbis spirorbis, Müll. (3) (5)
      Planorbis (Coretus) corneus (Linn.) (3) (5)
      Planorbis (Gyraulus) albus Müll.
      Planorbis (Gyraulus) glaber, Jeff. (3)
      Planorbis (Gyraulus) crista (Linn.) [=*nautileus*
        Linn.] (3)

Planorbis (Bathyomphalus) contortus (Linn.) (3)
Planorbis (Hippeutis) fontanus (Lightf.) [=*nitidus* Müller of Jeffreys] (3)
Segmentina nitida (Müll.) [=*lineata* Walker] (3)

PHYSIDAE
    Physa fontinalis (Linn.) (3) (5)
    Aplecta hypnorum (Linn.) (3)

STYLOMMATOPHORA

TESTACELLIDAE
    Testacella haliotidea, Drap.

LIMACIDAE
    Limax (Heynemannia) maximus, Linn.
    Limax (Lehmannia) flavus, Linn.
    Agriolimax agrestis (Linn.) (3)
    Agriolimax laevis (Müll.) (3)
    Milax sowerbyi (Fér.)
    Milax gagates (Drap.)

ZONITIDAE
    Vitrina pellucida (Müll.)
    Vitrea crystallina (Müll.) (3)
    Vitrea (Polita) lucida (Drap.) [=*draparnaldi* Beck]
    Vitrea (Polita) cellaria (Müll.) (3)
    Vitrea (Polita) rogersi, B. B. Woodward [=*glabra* Auctt. & *helvetica* Auctt.]
    Vitrea (Polita) alliaria (Miller)
    Vitrea (Polita) nitidula (Drap.) (3)
    Vitrea (Polita) pura (Ald.)
    Vitrea (Polita) radiatula (Ald.) (3)
    Zonitoides nitidus (Müll.) (3)
    Zonitoides excavatus (Bean)
    Euconulus fulvus (Müll.) (3)

ARIONIDAE
    Arion ater (Linn.)
    Arion intermedius, Normand [=*minimus* Simroth]
    Arion hortensis, Fér.
    Arion fasciatus, Nilsson [=*bourguignati* Mabille]

ENDODONTIDAE
  Punctum pygmaeum (Drap.) (3)
  Sphyradium edentulum (Drap.) [=*Vertigo edentula*]
    (3)
  Pyramidula (*Gonyodiscus*) *rotundata* (Müll.) (3)
HELICIDAE
  Helicella itala (Linn.) [=*ericetorum* Müll.] (3)
  Helicella (*Heliomanes*) *virgata* (Da C.) (3)
  Helicella (*Candidula*) *caperata* (Mont.) (3)
  Helicella (*Theba*) *cantiana* (Mont.)
  Hygromia (*Fruticicola*) *granulata* (Ald.) [=*sericea*
    Auctt. *non* Drap.]
  Hygromia (*Fruticicola*) *hispida* (Linn.) [=*concinna*
    Jeff.] (2) (3)
  Hygromia (*Fruticicola*) *rufescens* (Penn.) (3)
  Acanthinula aculeata (Müll.) (3)
  Vallonia pulchella (Müll.) (3) (5)
  Vallonia cristata (Müll.)
  Helicigona lapicida (Linn.) (3)
  Helicigona (*Arianta*) *arbustorum* (Linn.) (3)
  Helix (*Helicogena*) *aspersa*, Müll.
  Helix (*Helicogena*) *pomatia*, Linn.
  Helix (*Cepaea*) *nemoralis*, Linn. (3)
  Helix (*Cepaea*) *hortensis*, Müll.
ENIDAE
  Ena obscura (Müll.) (3)
STENOGYRIDAE
  Cochlicopa lubrica (Müll.) (3)
  Caecilioides acicula (Müll.) (3)
VERTIGINIDAE
  Jaminia muscorum (Linn.) [=*marginata* Drap.] (3)
  Jaminia (*Lauria*) *cylindracea* (Da C.) [=*umbilicata*
    Drap.] (3)
  Vertigo (*Alaea*) *antivertigo* (Drap.) (3)
  Vertigo (*Alaea*) *pygmaea* (Drap.) (3)
  Vertigo (*Alaea*) *moulinsiana* (Dup.) (3)

CLAUSILIIDAE
>   *Balea perversa* (Linn.) (3)
>   *Clausilia (Marpessa) laminata* (Mont.)
>   *Clausilia (Pirostoma) bidentata* (Ström.) [=*per-
>   versa* Pulteney=*rugosa* Drap.] (3)

SUCCINEIDAE
>   *Succinea putris* (Linn.) (2) (3) (5)
>   *Succinea elegans,* Risso (3) (5)

PELECYPODA

EULAMELLIBRANCHIATA

DREISSENSIIDAE
>   *Dreissensia polymorpha* (Pall.)

UNIONIDAE
>   *Unio pictorum* (Linn.) (3)
>   *Unio tumidus,* Retz (1) (3 ?)
>   *Anodonta cygnaea* (Linn.) (3 ?)

SPHAERIIDAE
>   *Sphaerium rivicola* (Leach)
>   *Sphaerium corneum* (Linn.) (3) (5)
>   *Sphaerium lacustre* (Müll.) (3)
>   *Pisidium amnicum* (Müll.) (2 ?) (3) (4) (5)
>   *Pisidium henslowianum* (Shepp.) (3)
>   *Pisidium subtruncatum,* Malm [=*fontinale* Jeff.] (3)
>   *Pisidium pulchellum,* Jenyns (3)
>   *Pisidium pusillum* (Gmel.) [=*fontinale* Drap. of
>   Continental authors] (3) (5)
>   *Pisidium nitidum,* Jenyns (3) (5)
>   *Pisidium obtusale,* Pfeiff. (5)

Compared with the British Isles as a whole the incidence
of species of non-marine Mollusca in Cambridgeshire is shown
by the following table.

	Cambridge-shire Species	British Species		Cambridge-shire Species	British Species
GASTEROPODA.			*Geomalacus*	0	1
Neritina	1	1	*Punctum*	1	1
Pomatias	1	1	*Sphyradium* ...	1	1
Acicula............	0	1	*Pyramidula*......	1	2
Assemanea	0	1	*Helicella*	4	6
Bithynia	2	2	*Hygromia*	3	5
Pseudamnicola .	0	1	*Acanthinula*......	1	2
Paludestrina ...	2	5	*Vallonia*	2	3
Vivipara	2	2	*Helicodonta*......	0	1
Valvata	2	2	*Helicigona*	2	2
Carychium	1	1	*Helix*	4	5
Phytia............	0	1	*Ena*	1	2
Ovatella	0	1	*Cochlicopa*	1	1
Ancylus	1	1	*Azeca*	0	1
Acroloxus.........	1	1	*Caecilioides* ...	1	1
Limnaea	5	8	*Jaminia*	2	4
Amphipeplea ...	1	1	*Vertigo*............	3	9
Planorbis.........	10	10	*Balea*	1	1
Segmentina	1	1	*Clausilia*	2	4
Physa	1	1	*Succinea*	2	3
Aplecta............	1	1			
Testacella	1	3	PELECYPODA.		
Limax	2	3	*Dreissensia*	1	1
Agriolimax	2	2	*Unio*...............	2	3
Milax	2	2	*Anodonta*	1	1
Vitrina......... ...	1	1	*Sphaerium*	3	4
Vitrea	8	8	*Pisidium*	7	8
Zonitoides	2	2			
Euconulus	1	1		101	143
Arion	4	6			

Notes on certain species.

Bithynia tentaculata. A bulloid monstrosity of this species was found in the Cam at Cambridge by the Rev. E. S. Dewick in 1886 and was described and figured by Mr E. A. Smith[1].

Pomatias elegans is certainly quite rare. The only record of its capture in the living state in Cambridgeshire is given

[1] *Journ. Conch.*, v., 1888, p. 315.

by Jenyns[1]. This is based on a single individual found at
Linton about 1830. He also mentions having found empty
shells at Reche Chalk Pits in 1827. Possibly these were
evidence of recently living individuals. Shells have been
found by Mrs Hughes in the Barrington Gravel and also in
pre-Roman tumuli, sometimes numerously, at various places,
including Upper Hare Park near Six Mile Bottom, Swaffham
Prior, Allington Hill and Chesterford. On the whole the
evidence is that this pretty form is in Cambridgeshire, as in
several other counties, becoming less common than formerly.

Amphipeplea glutinosa, a species inclined to sudden ap-
pearances and disappearances, is recorded by Jenyns[2] as
occurring in 1822, previously to which it had not been
recorded in the county, in pools in gravel pits at Bottisham.
It was exceedingly numerous in the above year and persisted
in diminishing numbers through the next three years, after
which it vanished entirely.

The same part of the county is recorded by the Rev. A. H.
Cooke[3] as that wherein *Aplecta hypnorum* was first noticed in
Cambridgeshire. During the wet February of 1852 a quite
small puddle in Bottisham Park was crowded with this species,
which completely and permanently disappeared a few days
afterwards.

Helicella itala. It is of interest that Pennant[4] mentions
"woods in Cambridgeshire" as a locality for this form.

Helicigona lapicida was found by Jenyns[5] in 1822 at
Swaffham Bulbeck and has since been captured at Fen Ditton
and as empty shells at Cherryhinton.

Helix aspersa. A dead sinistral example has been found
in the chalk pit at Cherryhinton.

"*Helix pomatia* lives locally at Maggots Mount near Little
Shelford. This, with the exception of its occurrence at

---

[1] *Observations in Natural History*, 1846, p. 322.
[2] *loc. cit.*, p. 318.
[3] *Cambridge Natural History*, Vol. v., p. 46.
[4] *British Zoology*, Vol. iv., 1812, p. 302.   [5] *loc. cit.*, p. 321.

Woodford in Northamptonshire (where, however, it is supposed to have been introduced), is its most northerly record in these islands. It is not a thriving inhabitant, and attempts to plant a colony at Madingley failed[1]."

Cochlicopa lubrica. The writer is indebted to the Rev. A. H. Cooke for an account of the appearance in 1878 of numerous well-developed examples of this species in the interstices of the paving stones of the footway between King's Parade and the University Library, in which unlikely situation they were pointed out to him by Mr F. J. H. Jenkinson, University Librarian.

Vertigo moulinsiana was discovered in Wicken Fen in 1897 by Mr J. R. Le B. Tomlin and has been found subsequently by other searchers at the same place. As a post-tertiary form it occurs in the Barnwell and Grantchester Gravels as well as in other parts of England. Besides the Wicken record of this species in the living state Hitchin and Broxbourne in Hertfordshire, Eastleigh in Hampshire, Morden in Dorsetshire, and West Galway are the only places in which it is known alive in the British Isles. At Wicken it is found on the tall grasses and other plants in summer and autumn, and Mr R. Standen[2] considers that the animal hibernates in hollows of the dead stems. He also calls attention to the suggestion by Jeffreys[3] made many years ago that this species would very likely be found still existing in the fen-land. *V. antivertigo* also occurs at Wicken.

Anodonta cygnaea. The specimen from the Cam at Cambridge in the University Museum of Zoology is the probable variety *anatina*, established by Linnaeus as a distinct species and as such regarded by many subsequent authors. The type form also occurs in the neighbourhood of Cambridge, but it is uncertain which is the commoner.

[1] Article "Molluscs" in *Victoria County Histories*, Vol. *Cambridgeshire*, by B. B. Woodward and H. H. Brindley.

[2] *Journ. Conch.*, IX., 1898, p. 217.

[3] *British Conchology*, 1862, I., p. 256.

132   *The Mollusca of Cambridgeshire*

Unio pictorum and *U. tumidus.* Mrs Hughes informs me
of the occurrence of these species at Barrington. Their distribution is probably local in the county.

Sphaerium corneum. In June 1901 Mr A. E. Shipley
found that a choking of the inflow of the Bathing Pool in
Christ's College Garden was caused by clustered masses of
this form. Jenyns[1] mentioned "a small globular variety" as
common in the Fens.

The *Pisidia* are well represented in Cambridgeshire, and as
long ago as 1832 were with *Sphaerium* the subject of a mono-
graph by the Rev. L. Jenyns[2]. His nomenclature was revised
by Jeffreys[3], but our knowledge of the line between species and
variety in *Pisidium* is still unsatisfactory. The writer has here
followed the provisional nomenclature of Mr B. B. Woodward[4].
P. amnicum is common and is found more often in moving
water than in pools. Jenyns recorded that Prof. Henslow
found several varieties in the neighbourhood of Cambridge.
The *P. obtusale* of Pfeiffer, either a distinct species or a
variety of *P. pusillum*, was found by Jenyns chiefly in
stagnant waters and often in the same spots as the Physidae.
P. nitidum occurs in clear waters, while a common form is
the *P. pulchellum* of Jenyns. *P. henslowianum*, Jenyns, was
first found by Prof. Henslow in ditches joining the Cam at
and below Cambridge. *P. amnicum* and *P. pusillum* have
also been found in the north of the county.

Pleistocene Species.

We may now turn to the occurrence of mollusca in the
post-tertiary deposits of the district. In addition to the
forms they contain which are also living at the present time
the Pleistocene beds yield a number of species which are now

[1] "British Species of Cyclas and Pisidium," *Trans. Camb. Phil. Soc.*,
1832, p. 289.
[2] *loc. cit. supra.*   [3] *loc. cit.*, pp. 19—28.
[4] "List of British Non-Marine Mollusca," *Journ. Conch.*, x., 1903,
p. 353.

extinct within the county or in the British Isles altogether. These forms have been recorded by several authors, the most recent and important papers on the subject being by Mrs Hughes[1] and Mr B. B. Woodward[2]. To these the reader is referred for details and conclusions. The following notes and the incidence of still living species in the Pleistocene beds already given in this article are extracted from these sources, from Mr F. Cowper Reid's work on the geology of the county[3], from the article on Mollusca in the "Victoria History" of the county and from malacological papers in various journals.

The marine shell-bearing gravels of the March district were laid down in the estuaries of the several streams which flowed into the Wash when its southern shores extended through what is now the north of Cambridgeshire. With about 24 species of marine Gasteropoda and about 20 species of marine Pelecypoda (all being forms extant at the present day) the March Gravels contain *Bithynia tentaculata, Valvata piscinalis, Unio tumidus* and *Corbicula fluminalis?*, but no Euthyneurous Pulmonata. In the neighbourhood of Cambridge, as at Girton and other places, there are gravels which are believed to have been deposited by affluents of the ancient rivers once debouching into the Wash, which then extended further south. In these gravels there are no marine shells but they contain *Hygromia hispida, Jaminia muscorum?, Succinea putris* and *Pisidium amnicum?*.

After the deposition of these gravels the present river system of the county became established, though there is some dispute as to the stages by which the ancient system passed into it. The gravels of the Ouse Valley at Somersham are among those laid down in the earlier period of the present river system and in them we find *Cardium edule* along with the estuarine form *Corbicula fluminalis*. Later than the Somersham beds are the gravels of Barrington, Barnwell and

[1] *loc. cit.*   [2] *Proc. Geol. Assoc.*, x., 1888, p. 355.
[3] *The Geology of Cambridgeshire.*

Grantchester, in which no marine species occur but which are very rich in freshwater and terrestrial forms. Most of those at present known are still inhabitants of Cambridgeshire, while some have apparently died out, though still living in the bordering or in more distant counties. There are also about half-a-dozen which are no longer British. On the other hand there are two or three species not found in the gravels which are living inhabitants of the county.

Species from the local gravels which do not now live in Cambridgeshire or at all events in the neighbourhood of Cambridge though still to be found within the British Isles are:—

Limax arborum, Bouch.-Chant., which is widely distributed on the Continent and in the British Isles, does not appear to have been found alive in Cambridgeshire, though it occurs in the bordering county of Northampton. It has also been found in a Holocene deposit containing unglazed Romano-British pottery at Harlton[1].

Acanthinula lamellata (Jeff.), which now lives in England as far south as the neighbourhood of Reading and also occurs in Scotland, northern Ireland, Sweden and northern Germany. This species has trended northward, at the same time becoming rare or dying out in the south[2].

Helicodonta obvoluta (Müll.), the "cheese snail," which is fairly common throughout Central Europe but is apparently dying out in England, wherein it is confined to beech woods on the chalk hills of Hants, Sussex and Surrey. The single example which has been obtained from the Grantchester Gravels is the only evidence of the range of this species north of the above counties.

Ena montana (Drap.) occurs locally but fairly numerously in the south and west of England, but in East Anglia seems to have been recorded only from West Suffolk. On the Continent it ranges from Silesia to the Pyrenees.

[1] R. A. Bullen, *Proc. Malacol. Soc. London*, v., 1903, p. 318.
[2] *Vide* B. B. Woodward, *Proc. Malacol. Soc. Lond.*, v., 1903, p. 11, for details of the present range of this species in England.

Vertigo minutissima (Hartm.) now ranges from Finland to Corsica and the Vosges Mountains, while it has been found living in scattered localities in Great Britain from Fifeshire to the Isle of Wight, but not in East Anglia.

Vertigo pusilla, Müll., concerning whose occurrence in the local gravels there is some doubt, owing to the loss of the single specimen recorded, is found from Perth to north Devon and in the north and west of Ireland. It occurs on the borders of Cambridge in Northamptonshire and West Norfolk as well as in East Suffolk, so that further search will very likely prove it to be still a living inhabitant of Cambridgeshire.

Vertigo angustior, Jeff., is a rare species which is found in Central Europe and Scandinavia, and in England has been reported from Westmoreland, the neighbourhood of Swansea, Tenby, Bristol and near London. It has been found in the Holocene bed at Harlton mentioned above, so it appears to have lived in Cambridgeshire in the Roman period.

Succinea oblonga, Drap., is not common in the British Isles, but exists near the coast in Ireland, Scotland, Wales and in Devonshire.

The following species which no longer live in the British Isles are recorded from the Cambridge district gravels.

Paludestrina marginata, Mich., lives in France at the present day.

Pyramidula ruderata, Studer, is now found in western Asia and apparently in Japan, as well as on the continent of Europe.

Eulota fruticum (Müll.) now ranges from northern Sweden over the greater part of Europe as well as through Siberia. The disappearance of the last two forms from Britain and their present day wide distribution elsewhere show how much we have to learn regarding the factors which determine the continued existence of a species in any region.

Clausilia pumila (Ziegler MS.), Pfeiffer, occurs not nearer the British Isles than western Germany.

Corbicula fluminalis, Müll., one of the Cyrenidae, which occurs doubtfully in the March shell-bearing gravels, now lives no further north than Asia Minor and Egypt. It seems to be a species which is dying out in the north, for in England it is found in the Norwich Crag and in the gravels not only of Cambridgeshire but in those of the Humber and Thames and of certain rivers of France, Belgium and Siberia.

Unio littoralis, Lam., still lives in France, the Iberian Peninsula and North Africa.

The variety of *Unio pictorum* described as *limosa* by Nilsson is now found alive in rivers in northern France.

Turning from the species which occur only in the Gravels, it is pointed out in Mrs Hughes's paper that near Cambridge there are now living certain common forms which hitherto have not been found in the Pleistocene deposits and therefore perhaps did not enter England till after these were laid down. These are *Helicella cantiana*, *Hygromia rufescens* and *Helix aspersa*, all found in abundance in the district at present. The question of whether the last named was imported into England has been much discussed, but the facts given in a recent paper by Messrs A. S. Kennard and B. B. Woodward[1] leave no doubt that though commerce has probably extended its range this species entered the British Isles without the help of man. The authors conclude that it came from south-western Europe over lands now submerged in the Atlantic in the same manner as *Geomalacus maculosus* entered Ireland, *Hygromia montivaga* Cornwall and *Helix pisana* South Wales, Cornwall and the Channel Islands. *Helix aspersa* is not known as a fossil on the continent of Europe, but it occurs in Pleistocene deposits in Algeria, while in England it is found in Neolithic beds at Hastings, in beds at least pre-Roman at Harlyn Bay in Cornwall, in hill wash at St Catherine's Down, I. of Wight, and in Holocene deposits at Clifden Hempden near Oxford.

Helix pomatia, which lives locally near Shelford, is not known in the Gravels and Mrs Hughes failed to find it with *H. aspersa* in the Roman pits. In this connection may be

[1] *Proc. Cotteswold Naturalists' Field Club*, xiv., 1903, p. 199.

noted that the discovery of the species in the fossil state near Reigate dispels the often stated belief that it was a Roman importation[1]. *Helicigona arbustorum* and *Helix nemoralis* are quite common in the neighbourhood of Cambridge, and, as has been already stated, in particular near the river. Both species are found in the local Gravels, but *H. nemoralis* not numerously, so it is possible that it is increasing in the district.

The suggestion has been made that further search may very possibly add to the list of Cambridgeshire species. If we consider the table below, in which the species given are all living in one or more of the bordering counties[2], while two or three occur also in the Pleistocene Gravels of the Cambridge district, the opinion advanced seems to have reasonable support.

	Pleistocene Gravels of Cambridge district	North Essex	Hertford	West Norfolk	Bedford	Northampton	South Lincoln
Testacella scutulum	...	...	...	+	...	...	...
Limax arborum	+	...	...	...	...	+	...
Arion subfuscus	...	...	...	...	...	...	+
Pyramidula rupestris	...	...	+	...	...	+	...
Hygromia fusca......................	...	...	...	...	...	+	...
Ena montana	+	...	...	...	...	...	+
Azeca tridens	...	+	...	...	+	+	...
Vertigo substriata	...	...	...	+	...	...	...
Vertigo pusilla	?	...	...	+	...	+	...
Clausilia biplicata	...	...	+	...	...	...	...
Clausilia rolphii	...	...	...	...	...	+	...
Pisidium gassiesianum...............	...	...	+	...	+	+	...

[1] *Vide* R. A. Buller, *Proc. Malacol. Soc. Lond.*, III., 1899, p. 326, for details and discussion of this subject.

[2] *Vide* L. E. Adams, "Census of the British Land and Freshwater Mollusca," *Journ. Conch.*, x., no. 7, July, 1902.

Huntingdon and West Suffolk, the bordering counties not included in this table, do not possess any species not found also in Cambridgeshire.

There are in the Fenland several shell-bearing deposits more recent than the Pleistocene Gravels and representing the period when the land to the east and north of Cambridge was still covered by brackish and tidal waters, dotted with more or less isolated patches of dry land, of which the Isle of Ely as it then was formed one of the largest. In the formation known as Fen silt there occur shells of *Mytilus edulis, Ostrea edulis, Cardium edule* and *Scrobicularia plana.* The last-named is so frequent in one bed as to have obtained for it the title "Scrobicularia Clay." *Pisidium amnicum* also occurs in the Fen silt, while another Fen deposit, the Shell Marl of Burwell and other places, contains about 20 species of non-marine Gasteropoda and about 5 of non-marine Pelecypoda, all of which are living in the county at the present time. *Paludestrina ventrosa,* which is not known alive in the county, occurs near Littleport in the Scrobicularia Clay.

The public collections of mollusca obtained from within the county are those in the Samuel Smith and other cabinets in the Wisbech Museum, the Cambridge Natural History Society's County Type Collection in the University Museum of Zoology, while the general and MacAndrew Collections in the latter also contain many specimens from the Cambridge neighbourhood. The Sedgwick Museum of Geology has most of the species found in the local Gravels.

In compiling this article the writer is indebted to Mrs McKenny Hughes, the Rev. A. H. Cooke (King's College), Head Master of Aldenham School, and Mr F. C. Morgan, B.A. (Trinity College), for much information as to occurrences and localities, and also to Mr B. B. Woodward, F.L.S., of the British Museum, for most valuable assistance on nomenclature, identifications and references to the literature of the subject.

THE INSECTS OF CAMBRIDGESHIRE.

INTRODUCTION.

By Wm Farren.

CAMBRIDGESHIRE, although small, possesses a remarkably rich and varied Entomological fauna, due to a considerable variety of country being included in the long straggling shape of the county.

Of real fen-land very little remains; the whole of the northern part of the county, the centre of that large tract of Fens which extends into north Huntingdonshire, Lincolnshire, Norfolk and Suffolk, has been reclaimed, and is now, with the exception of small isolated parts, agricultural land of a somewhat uninteresting character, at least from the entomologist's point of view. The Fens of Horningsea, Bottisham, Swaffham, and Burwell, immediately north of Cambridge, have also been reclaimed in the last thirty or forty years, and although a semi-wild state is maintained in parts where turf-digging is carried on, the greater part is cultivated land, and but few of the interesting insects which were to be found there in the middle of last century remain. About 300 acres of sedge fen at Wicken alone survives in something of its original state; a succession of bad hay seasons has however raised the value of sedge and such rough fodder as the Fen produces, and a considerable amount of cutting has been done, to facilitate which a lot of bushes have been "stubbed" up. Against this stands the fact that several naturalists have purchased portions of the Fen with the object of preserving it in its original state, and this, with the almost insurmountable difficulties to be encountered in draining, makes it very unlikely that any further changes will take place in the near future; Wicken contains much of the old Fen flora and insect fauna, but bears a poor resemblance to what the more northern Fenland, with its vast meres and reed beds, was like. Everything is so condensed, and the collecting of insects made so easy by professional attendance near at hand, that it might more aptly be termed an entomological nursery!

On the extreme west of the county the Fens are lost a few miles south of the river Ouse, and the boundary is dove-tailed into those of south Huntingdonshire and Bedfordshire, adding to the Cambridgeshire fauna a rich share of the produce of the woods of those counties. Here are the woods and heaths near Gamlingay, so often mentioned by Jenyns and the old collectors. In the south commences that extensive line of chalk-hills from north Hertfordshire and Essex, which running in a north-easterly direction along the edge of the Fen country immediately north of Cambridge to Burwell, Fordham and Newmarket, covers nearly the whole of the south and south-east of the county, and is very rich in chalk-frequenting insects. Five miles north of Newmarket, separated from the lower Fen country by some comparatively high land where the Fens are bounded by the "Breck sands" of north-west Suffolk, is Chippenham Fen; this has been preserved for game for a number of years, and the Fen is surrounded and crossed by plantations of mixed trees which vary the Fen fauna considerably with woodland species. Here occur many of the Wicken insects, and other Fen species not to be found there, such as *Plusia chryson* and *Bankia argentula* among the Lepidoptera.

East and north of Chippenham we get a narrow piece of the "Breck" in Cambridgeshire, sufficient to add to our Lepidopterous fauna *Dianthoecia irregularis*, *Agrophila trabealis*, *Acidalia rubiginata*, and other species peculiar to that district, including some which are otherwise coast species such as *Agrotis vestigalis*, and *Lita semidicandrella*.

South of Cambridge are some productive reed-beds in the valley of the upper Cam and its tributaries.

Thus it will be seen that, although the whole of the northern part of the county is Fen land, in the south it is varied with a lot of wooded country west of Cambridge, woods and heaths in the extreme west, chalk hills in the south and south-east, and the narrow strip of "Breck sand" above Newmarket.

Apart from the LEPIDOPTERA very little is known about the insects of the county; several orders have been collected to some extent at Wicken and Chippenham, but very little else has been done, or at any rate recorded. The late Rev. L. Jenyns (afterwards Blomefield), who lived at Bottisham in the early half of last century, left to the Cambridge Philosophical Society a manuscript catalogue of the insects of Cambridgeshire, together with his collections of insects, which include the actual specimens referred to in the catalogue, except in the case of the species whose records were supplied to him by other workers. This catalogue has been largely drawn upon for the following articles, and in some orders, such as ORTHOPTERA, NEUROPTERA, HYMENOPTERA, and HEMIPTERA, it forms almost the whole available material. With the exception of scattered records in the various Entomological Magazines, the only published matter appears to be the list of "Lepidoptera of the Fens" in *Fenland Past and Present*, 1878; this list is valuable as it includes the records of Lord Walsingham and Mr Wm Warren and, more important still, of the late Mr F. Bond and Mr A. Thurnall. In addition to the Jenyns collection in the museum there is that of the late Prof. Babington, which is especially rich in COLEOPTERA, and another large collection of local LEPIDOPTERA which contains long series of all the extinct Fen species.

Of the following articles and lists, those on ORTHOPTERA, NEUROPTERA, HEMIPTERA, and DIPTERA are copies of those prepared for the forthcoming Volume on Cambridgeshire, of the Victoria County History, and are here inserted by the courtesy of the Editors of that work.

THE ORTHOPTERA OF CAMBRIDGESHIRE.

By MALCOLM BURR, B.A., F.L.S., F.Z.S., F.E.S.

THE following account of the Orthoptera of Cambridgeshire is based upon the collection made by the late Rev. L. Jenyns, and most of the specimens referred to are to be found in the collection, left by him to the Cambridgeshire Philosophical Society. A few other localities have been added, based upon recent captures recorded in the magazines.

DERMATOPTERA.
Family FORFICULIDAE.

Labia minor, L. Common in the summer, often seen on the wing in company with *Staphylinidae*, over flower beds and dungheaps.

Forficula auricularia, L. Abundant everywhere (Jenyns). Var. *forcipata*, Steph. "Not uncommon at Bottisham in harvest time, in sheaves of wheat" (Jenyns).

Apterygida media, Hagenb. "Cambridge, Prof. C. C. Babington." This is an exceedingly rare species; it has, in fact, been only recorded on two other occasions in Great Britain; Westwood took it at Ashford, and Mr James Edwards has taken it near Norwich.

DICTYOPTERA.
Family BLATTIDAE.

Stylopyga orientalis, L. Swarms in houses.

(*Periplaneta Australasiae*, Fabr., another introduced species, is plentiful in the Palm-houses and warm potting-sheds in the Cambridge Botanic Gardens; it is also very probable that some at least of the three British *Ectobiidae* occur, but records of captured Orthoptera in the county are few and far between.)

EUORTHOPTERA.

ACRIDIODEA.

Family TRUXALIDAE.

Mecostethus grossus, L. "Cambridge" (C. W. Dale). Abundant in the fens round Ely; also in Burwell and Swaffham fens, Gamlingay bogs, etc. (Jenyns).

Stenobothrus lineatus, Panz.

Omocestus viridulus, Linn. Common in meadows, Burwell, etc. (Jenyns); common at Wicken Fen (Porritt).

O. rufipes, Zett. Common at Wicken Fen (Porritt).

Stauroderus bicolor (Charp). An abundant species in most places, Burwell, etc. (Jenyns); abundant at Wicken (Porritt). This species is exceedingly variable in colour, and nearly as common as it is variable.

Chorthippus elegans (Charp). "Bottisham Fen" (Jenyns); also Whittlesea Mere and Wicken Fen. This species occurs in a few scattered localities, but is usually numerous where it does occur. Not uncommon at Wicken (Porritt).

Ch. parallelus (Zett). Abundant throughout Britain. "On rushes and fen-ditches," Thorney (Jenyns).

Gomphocerus maculatus, Thunb. "Devil's Ditch and Newmarket Heath, Gamlingay, Wilbraham Temple" (Jenyns).

Family OEDIPODIDAE.

Pachytylus danicus, Linn., a so-called *Locusta migratoria*, is recorded from Duxford in September 1847, and also near Wisbech and Fulbourn (Jenyns). It is probably referable to this species.

Family TETTIGIDAE.

Tettix subulatus, L. "Coe Fen near Cambridge in April; not uncommon" (Jenyns).

T. bipunctatus, L. Bottisham (Jenyns). Probably common everywhere.

LOCUSTODEA.

Family PHANEROPTERIDAE.

Leptophyes punctatissima, Bosc. Gamlingay Wood on trees in August; Bottisham Park on poplars (Jenyns); common at Wicken (Porritt).

Family MECONEMIDAE.

Meconema varium, Fabr. "The Backs" (D. S.). Common in the autumn near Bottisham (Jenyns); beaten out of trees in plenty at Chippenham (Porritt).

Family CONOCEPHALIDAE.

Xiphidium dorsale, Latr. Burwell and Bottisham Fens near Cambridge (Jenyns); in abundance on Chippenham Fen, also on Wicken Fen, but less commonly (Porritt). This species is very locally distributed in England. "Another unnamed species allied to this" is referred to by Jenyns, but its identity is doubtful.

Family LOCUSTIDAE.

Locusta viridissima, L. Very common in the fens and in hedges, etc., in many places (Jenyns). This fine species appears to be less common than formerly.

Family DECTICIDAE.

Thamnotrizon cinereus, L. Gamlingay, near Wharesby Park, in August among plants and grass by the roadsides, Shudy-camps (Jenyns).

GRYLLODEA.

Family GRYLLIDAE.

Gryllus domesticus, L. Common in houses (Jenyns).

Family GRYLLOTALPIDAE.

Gryllotalpa gryllotalpa, L. In plenty at Fulbourn; in the park at Bottisham near the canal (Jenyns).

THE NEUROPTERA OF CAMBRIDGESHIRE.

By Kenneth J. Morton, F.E.S.

The following lists of Neuroptera, in the broad sense, occurring in Cambridgeshire, are based on a MS. Catalogue by the late Rev. L. Jenyns[1], and on collections made in the county by Mr Geo. T. Porritt and myself, the authority in each case being indicated by the respective initials. In Mr Jenyns' catalogue the names given are those of Stephens; accordingly, in his case, the accuracy of the records depends on his having correctly identified Stephens' species as represented by the latter author's types determined by Mr McLachlan.

There seems to be no good ground however for concluding that any of the species here recorded have not occurred in the county.

TRICHOPTERA.

Phryganea grandis, L.   Near Ely (J).

,,   *striata*, L.   Near Ely (J); Thorney (M).

,,   *varia*, F.   Cambridge and elsewhere (J); Wicken (P).

[1] Jenyns in his MS. Catalogue gives a long list of Bird-lice, Mallophaga, under the heading 'Anoplura' with the following note:—"The list of species of *Anoplura* taken in Cambridgeshire has reference in each case to the *Monographia Anoplurorum Britanniae*, published in 1842 by Mr Henry Denny of Leeds; to whom all the specimens were sent for examination and comparison at the time when he was collecting materials for his work. All the species were named by him. This part of the collection was afterwards given to the Entomological Society of London,—and is therefore not to be found in the cabinet containing the other orders of Cambridgeshire Insects. (L. J.)"

Agrypnia pagetana, Curt.   Near Thorney (J and M);
Wicken (P).
Colpotaulius incisus, Curt.   Bottisham and Ely (J); Thorney
(M); Wicken (P).
Grammotaulius nitidus, Müller.   Thorney (M); Wicken (P).
   „   *atomarius*, F.   Bottisham and Ely (J);
                Thorney (M).
Glyphotaelius pellucidus, Retz.   Bottisham and Ely (J);
Thorney (M); Wicken (P).
Limnophilus rhombicus, L.   Bottisham and Ely (J); Wicken
         (P).
   „   *flavicornis*, F.   Bottisham and Ely (J); Wicken
         (P).
   „   *marmoratus*, Curt.   Not uncommon.
   „   *lunatus*, Curt.   Ely, Cambridge and Bottis-
         ham (J).
   „   *vittatus*, F.   Ely, Cambridge and Bottisham (J).
   „   *affinis*, Curt.   Ely, Cambridge and Bottisham
         (J); Thorney (M).
   „   *auricula*, Curt.   Ely, Cambridge and Bottisham
         (J).
   „   *griseus*, L.   Ely, Cambridge and Bottisham (J).
   „   *bipunctatus*, Curt.   Ely, Cambridge and Bot-
         tisham (J).
   „   *hirsutus*, Pict.   Chippenham (P).
Anabolia nervosa, Curt.   Bottisham, &c. (J).
Stenophylax permistus, McL.
Halesus digitatus, Schnk.   Bottisham, &c. (J).
Chaetopteryx villosa, F.   "Cambridgeshire" (Stephens).
Notidobia ciliaris, L.   Bottisham (J).
Molanna angustata, Curt.   Bottisham (J); Thorney (M);
Wicken (P).
Leptocerus fulvus, Ramb.   Thorney (M).
   „   *senilis*, Burm.   Wicken (P).
   „   *aterrimus*, Steph.   Common.
   „   *cinereus*, Curt.   Bottisham (J); Wicken (P).

Mystacides longicornis, L.   Near Thorney (J); Wicken (P).
Triaenodes bicolor, Curt.   Bottisham (J); Wicken (P).
Erotesis baltica, McL.   Wicken and Chippenham (P).
Oecetis ochracea, Curt.   Thorney (M).
„   *furva*, Ramb.   Wicken (P).
„   *lacustris*, Pict.   Wicken (P).
Oxyethira costalis, Curt.   Thorney (M).
Hydropsyche pellucidula, Curt.   Bottisham (J).
Polycentropus flavomaculatus, Pict.   Bottisham (J).
Holocentropus picicornis, Steph.   Wilbraham (J); Thorney
(M); Wicken (P).
„   *dubius*, Ramb.   Wicken (P).
Tinodes waeneri, L.   Bottisham (J).
Agapetus comatus, Pict.   Bottisham (J).

NEUROPTERA-PLANIPENNIA.

SIALIDAE.

Sialis lutaria, Lin.   Common.

PANORPIDAE.

Panorpa communis, Lin.   Abundant (J).
„   *germanica*, Lin.   Common (J).

RHAPHIDIIDAE.

Raphidia xanthostigma, Schum.   Chippenham (P).

OSMYLIDAE.

Sisyra fuscata, F.   Bottisham (common J).

HEMEROBIIDAE.

Micromus variegatus, Fab.   Bottisham (J).
„   *paganus*, L.   Bottisham and Cambridge (J).
Hemerobius micans, Oliv.   Thorney (M).
„   *humuli*, L.   Near Bottisham (J).

Hemerobius lutescens, Steph.   Bottisham and Cambridge (J).
„   *marginatus*, Steph.   Chippenham (P).
„   *stigma*, Steph.   Bottisham and Cambridge (J).
„   *pini*, Steph.; near Bottisham (J).
„   *subnebulosus*, Steph.   Thorney (M).
„   *nervosus*, F.   Gamlingay (J).
Megalomus hirtus, L.   Near Bottisham (J).

CHRYSOPIDAE.

Chrysopa flava, Scop.   Chippenham (P): Thorney (M).
„   *vittata*, Wesm.   Chippenham (P).
„   *alba*, Lin.   Gamlingay (J); Chippenham (P).
„   *flavifrons*, Brauer.   Wicken (P).
„   *tenella*, Schn.   Chippenham (P).
„   *vulgaris*, Schn.   Bottisham (J).
„   *septempunctata*, Wesm.   Bottisham (J); Wicken
            (P); Thorney (M).
„   *aspersa*, Wesm.   Chippenham (P).
„   *ventralis*, Curt.   Bottisham (J).
„   *phyllochroma*, Wesm.   Thorney (M).

ODONATA.

Sympetrum striolatum, Charp.   Common.
„   *sanguineum*, Müll.   Not uncommon in the fens
            (Jenyns, M. and P.).
„   *scoticum*, Don.   Gamlingay and near Whittlesea
            (J); Thorney (M).
Libellula depressa, L.   Abundant in fens and elsewhere (J).
„   *quadrimaculata*, L.   Cambridge and Ely (J);
            Thorney (M).
Orthetrum coerulescens, Fabr.   Gamlingay (J); Chippenham
            Fen (P).
„   *cancellatum*, L.   Bottisham Park and occasionally
            elsewhere (J).

Anax imperator, Leach.   Newnham near Cambridge (J).
Brachytron pratense, Müll.   Bottisham (J); Thorney (M).
Aeschna juncea, L.   Less common than the next (J).
   ,,   *cyanea*, Müller.   Not uncommon (J); Thorney (M).
   ,,   *grandis*, L.   Common in the fens and in many places
           (J); confirmed (M and P).
   ,,   *isosceles*, Müller.   Swaffham Fen (J).
Calopteryx splendens, Harr.   Common (J).
Lestes dryas, Kby.   Thorney (M).
   ,,   *sponsa*, Hans.   Thorney (M).
Pyrrhosoma nymphula, Sulz.   Common.
   ,,   *tenellum*, Vill.   Gamlingay (J);   Wicken and
           Chippenham Fens (P).
Ischnura pumilio, Charp.   Gamlingay.
   ,,   *elegans*, Lind.   Common.
Agrion pulchellum, L.   Whittlesea and Thorney (J); Thorney
           (M).
   ,,   *puella*, L.   Occasionally (J); Thorney (M).
Enallagma cyathigerum, Charp.   Thorney (M); also taken by
    Jenyns, locality uncertain.

PERLIDAE.

Nemoura variegata, Oliv.   Common.
Leuctra geniculata, Steph.   Bottisham (J).

PSOCIDAE.

Psocus longicornis, F.   Bottisham (J).
Stenopsocus immaculatus, Steph.   Bottisham (J).
Caecilius flavidus, Curt.   Bottisham and elsewhere (J).

THE HEMIPTERA OF CAMBRIDGESHIRE.

By W. FARREN.

THE following lists of Hemiptera are mainly culled from Jenyns' MS. Catalogue. Mr Edward Saunders, F.L.S., F.E.S., has kindly revised the synonymy of the Heteroptera and added such Cambridgeshire records as appear in his work on the sub-order[1]. Such records have the recorder's name attached to the locality. All localities without a reference are from Jenyns' Catalogue. The list of Homoptera is entirely from Jenyns' Catalogue, and has been kindly revised by Mr James Edwards.

HEMIPTERA-HETEROPTERA.

GYMNOCERATA.

PENTATOMIDAE.

Corimelaena scarabaeoides, Lin. Devil's Ditch.
Eurygaster maura, Lin. Burwell Fen, 1825.
Sehirus bicolor, Lin. Cambridge, &c.
 „ *dubius*, Scop. Swaffham Bulbeck.
Gnathoconus albomarginatus, Fab. Swaffham Bulbeck.
Aelia acuminata, Linn. Devil's Ditch.
Eysarcoris melanocephalus, Fab. In profusion in Cherryhinton chalk-pits, June, 1828 (L. P. Garrons).
Pentatoma baccarum, Linn. Bottisham.
Strachia oleracea, Linn. Cambridge (Bond).
Piezodorus lituratus, Fab. Bottisham, and Newmarket Heath.
Tropicoris rufipes, Linn. Bottisham, and Reach Lode.
Picromerus bidens, Linn. Near Cambridge.
Podisus luridus, Fab. Bottisham, etc.; Wicken Fen (Blatch).
Asopus punctatus, Linn. Bottisham.
Acanthosoma haemorrhoidale, Linn.
 „ *dentatum*, De G. Both near Cambridge.

---

[1] *The Hemiptera-Heteroptera of the British Islands*, Saunders, 1892.

COREIDAE.

Myrmus miriformis, Fall. Two specimens taken *in copula* on the Devil's Ditch.

BERYTIDAE.

Berytus clavipes, Fab. Great Wilbraham.

LYGAEIDAE.

Heterogaster urticae, Fab. Bottisham.
Plociomerus fracticollis, Schill. Wicken (Champion).
Acompus rufipes, Wolf. Wicken Fen (Champion).
Eremocoris podagricus, Fab. Littlington (Power)—under dead leaves.
Scolopostethus pictus, Schill. Littlington, February (Power). Bottisham.
Drymus sylvaticus, Fab. Bottisham.
Gastrodes ferrugineus, Linn. Gamlingay, Alington Hill, etc.

TINGIDIDAE.

Piesma quadrata, Fieb.⎫
    „   *capitata,* Wolf.⎭ Swaffham Bulbeck.
Dictyonota crassicornis, Fall. Bottisham.
Derephysia foliacea, Fall. Devil's Ditch.
Monanthia ampliata, Fieb. Wicken (Champion); Ely (Billups).

ARADIDAE.

Aradus depressus, Fab. Bottisham Fen.

HYDROMETRIDAE.

Hydrometra stagnorum, Linn. Bottisham canal, etc.
Velia currens, Fab. Bottisham, etc.
Microvelia pygmœa, Duf. Wilbraham and Bottisham.
Gerris paludum, Fab. Wilbraham.
    „   *thoracica,* Schum. Bottisham Park canal.

REDUVIIDAE.

Ploiaria vagabunda, Linn.   On walls of damp out-buildings, Bottisham Hall.
Reduvius personatus, Linn.   In old houses—Ely, not uncommon.
Nabis lativentris, Boh.   Bottisham.
  „   *ferus*, Linn.   Newmarket Heath.

SALDIDAE.

Salda saltatoria, Linn.   Bottisham.
  „   *elegantula*, Fall.   Wicken Fen (Blatch).

CIMICIDAE.

Cimex lectularius, Linn.   Common bed-bug.
  „   *hirundinis*, Jen.   In house-martin's nests, sometimes very abundant, Swaffham Bulbeck.
  „   *pipistrelli*, Jen.   One specimen from the pipistrelle.
Anthocoris nemoralis, Fabr.   Bottisham.
  „   *sylvestris*, Linn.   Great Wilbraham, Reach, etc.
Microphysa pselaphiformis, Curt.   On faggot stacks in the wood-yard at Bottisham Hall farm.

CAPSIDAE.

Miris laevigatus, Linn.   Great Wilbraham, etc.
  „   *calcaratus*, Fall.   Burwell Fen.
Megaloceraea erratica, Linn.   Bottisham and Great Wilbraham.
Teratocoris antennatus, Boh.   Wicken Fen (Champion).
Leptopterna dolobrata, Linn.   Devil's Ditch.
Phytocoris ulmi, Linn.   Great Wilbraham.
  „   *varipes*, Boh.   Devil's Ditch.
Calocoris roseomaculatus, De G.   Devil's Ditch and Great Wilbraham.
Capsus laniarius, Linn.   Near Cambridge.
Rhopalotomus ater, Linn.   Swaffham Bulbeck, etc.

Campyloneura virgula, H. S.   Great Wilbraham.
Cyllocoris histrionicus, Linn.   Linton.
„   *flavonotatus*, Boh.   March (Billups).
Globiceps flavomaculatus, Fab.   Bottisham Park.
Orthotylus concolor, Kb.   March (Billups).
Oncotylus viridiflavus, Goeyh.   Devil's Ditch.
Harpocera thoracica, Fall.   Bottisham.
Plagiognathus arbustorum, Fab.   Linton.

CRYPTOCERATA.

NAUCORIDAE.

Naucoris cimicoides, Linn.   Very abundant in the pond at
Wilbraham Temple.

NEPIDAE.

Nepa cinerea, Linn.   Common.
Ranatra linearis, Linn.   In great plenty in the garden pond,
Wilbraham Temple.   Fens near Cambridge occasionally
(Jenyns).   Cambridge (Dale).

NOTONECTIDAE.

Notonecta glauca, Linn.   Common.
„   „   *var. furcata*, Fb.   Ponds near Cambridge.
„   „   *var. maculata*, Fb.   Bottisham Park canal.

CORIXIDAE.

Corixa geoffroyi, Leach.   Bottisham Park canal.
„   *striata*, Fieb.   With the last.
„   *fallenii*, Fieb.   Mildenhall drain.
„   *semistriata*, Fieb.   Cambridge fens (Douglas and
Scott).
„   *coleoptrata*, Fab.   Burwell Fen (Jenyns), Cambridge
fens (Douglas and Scott).
Sigara minutissima, Linn.   Cambridge (Billups).

HEMIPTERA-HOMOPTERA.

CICADINA.

MEMBRACIDAE.

Centrotus cornutus, Linn.   Gamlingay, Whitewood.
Issus coleoptratus, Geoffr.   Ely, Swaffham Bulbeck, etc.

CIXIIDAE.

Cixius nervosus, Linn.   Gamlingay Wood, Hildersham, Bottisham Fen.

CERCOPIDAE.

Triecphora vulnerata, Illiq.   Gamlingay in May (C. Darwin).
Philaenus spumarius, Linn.   Common.
  „   „   *var. lateralis*.   Near Ely.
  „   *lineatus*, Linn.   Bottisham.

BYTHOSCOPIDAE.

Macropsis lanio, Linn.   Bottisham Park, Gamlingay Wood.
Idiocerus varius, Fab.   Grunty Fen, near Ely.

TETTIGONIDAE.

Tettigonia viridis, Liv.   Fens near Ely, Great Wilbraham.
  „   „   *var. arundinis*, Germ.   Great Wilbraham, and Grunty Fen.

ACOCEPHALIDAE.

Acocephalus bifasciatus, Linn.   Bottisham, Gamlingay.
  „   *nervosus*, Schrk.   Bottisham.
Eupelix cuspidata, Fab.   Great Wilbraham.

PSYLLINA.

LIVIIDAE.

Livia juncorum, Latr.   Common on rushes in the Fens.

At the end of this list in Jenyns' Catalogue there is a postscript—"See a notice of the *Cochineal Insect* occurring in the Botanic Garden, Cambridge—(Loud. Mag. Nat. Hist. Vol. 2, p. 386)."

The notice referred to is an extract from "The Times" of Jan. 26th, 1829 : "Dr Gorman discovered a few weeks ago, the *Gròna sylvestris* among coffee plants, acacia, etc....... This is the kermes or gronilla of Spain...."

THE

COLEOPTERA OF CAMBRIDGESHIRE.

By Horace St J. K. Donisthorpe, F.Z.S., F.E.S.

Cambridgeshire, to the general collector, is somewhat uninteresting, being but a moderate sized county, very flat, and considerably cultivated, but is of course redeemed by the Fens, which support many rare and local insects. These are the remnants of very much larger marshy areas, which formerly occupied the whole of the district, and the species that haunt them have naturally been driven and condensed into this smaller space, and, indeed, may be compared to Hereward the Wake, holding it as their last stronghold. Although one or two species, which will be dealt with later, seem to have disappeared before drainage, as in the case of the Great Copper in the Lepidoptera, the majority of the fen-dwellers still survive at Wicken and other localities, and some are now only found there in Britain. The following notes only deal with the more local and rare species, and those of the purely fen-insects.

To commence with the *Geodephaga*, or Ground-beetles, the rare variety *consitus* of *Carabus monilis* has been taken at Wicken, and *Carabus granulatus* is common in the fens on paths, and under cut herbage, etc. Wollaston records a dozen specimens of the local *Calosoma inquisitor* as being once taken by a gentleman at Gamlingay. *Blethisa multi-punctata*, which occurs in marshy places, on the mud at the edges of pools and so on, is recorded from the Cambridgeshire Fens, as also the rare *Elaphrus uliginosus*. The handsome *Panageus crux-major* occurs sparingly at Wicken and other fens under sedge refuse; the very rare *Chlænius holosericeus* is recorded from Whittlesea Mere; from Fen Ditton by

S. Hanson in 1827, and was taken by Charles Darwin near Cambridge; later Dr Power took it in Burwell Fen, but it has not been taken recently. *Oodes helopioides*, another fen species, is not uncommon at times on the edges of turf-pits, etc. *Bradycellus placidus* is to be found under cut sedge at Wicken Fen, and is also recorded from Whittlesea Mere. The very local *Harpalus obscurus* appears to be almost confined to Cambridgeshire, occurring in its old locality at the foot of the Devil's Dyke, near Swaffham, though the writer once took two specimens at Abbotsbury in Dorsetshire. The two large species of *Amara*, *A. spinipes* and *A. convexiuscula*, may be swept in Wicken and Burwell Fens. The large *Sphodrus leucophthalmus*, which is generally taken in cellars, is recorded by the Rev. L. Jenyns as once having been taken at Bottisham Lode; it is also found in cellars in Cambridge. The two beautiful species of *Pterostichus*, *P. lepidus* and *P. dimidiatus*, have both been taken at Gamlingay. *Pterostichus aterrimus* is a species which seems to have quite disappeared before drainage, though it used to be abundant at Whittlesea Mere, Bottisham and other fens, basking in the sun on the soft mud at the edges of turf-pits. *Trechus rubens* was formerly common at Whittlesea Mere, and *T. discus* is recorded from Cambridgeshire. The very rare *Trechus rivularis* was recorded by Dawson from Whittlesea Mere in 1847: Mr A. J. Chitty, however, took two specimens under cut sedge at Wicken Fen in 1900, which proves that it is not extinct as was supposed. The graceful *Odacantha melanura* occurs not uncommonly among reeds in the fens. The two rare fen species, *Aëtophorus imperialis* and *Dromius longiceps*, are found in marshy places; the latter and the little *Dromius sigma* were formerly common in the sedge boats on the Cam.

Coming to the Water-beetles we find them, as might be expected, very well represented; space, however, will only permit us to deal with a few of the more prominent species. The rare *Haliplus mucronatus* is recorded from Cambridge, Wicken, and Soham. *Agabus abbreviatus* was formerly abundant in the Fens; *Rhantus exoletus* is not uncommon at

Wicken, where it often flies to light at night, and the very rare *Rhantus adspersus* has only occurred near Cambridge, where it was found in numbers in 1829; of late years it seems to have disappeared. The large *Dytiscus circumcinctus* and *D. dimidiatus* are not uncommon, at any rate in Wicken Fen, in ditches and peat holes; of the former the form of the female with smooth elytra is not very rare; the latter was supposed to have become exceedingly rare of late years; it must, however, have revived again, as it was by no means uncommon in 1900. *Hydaticus transversalis* occurs sparingly in Wicken Fen, and the other species, *H. seminiger*, is also recorded from the same locality. *Graphoderes cinereus* is another species which used to be taken in the Cambridgeshire Fens; it has not, however, been found for many years. *Gyrinus suffriani*, one of the rarer species of the "Whirligig" Beetles, is recorded from Wicken Fen.

In the Hydrophilidae the very large *Hydrophilus piceus* is common at Wicken Fen in pools and ditches, being very fond of getting under the leaves of the water-lily; and the smaller *Hydrocharis caraboides* has been recorded from Whittlesea Mere, Bottisham, etc. The very local *Spercheus emarginatus* was taken by Professor Babington in Burwell Fen. *Hydrochus carinatus*, which is only found in Huntingdonshire and Cambridgeshire, has been taken at Wicken, and other localities near Cambridge.

To mention a few of the rare species in the very large family *Staphylinidæ*, *Aleochara fumata* was taken by Mr Champion at Soham, and *Microglossa marginalis* was taken by Mr Crotch near Cambridge, the only British records for these two species. The very rare *Oxypoda verecunda* is recorded from Whittlesea Mere, and *Oxypoda pectita* from Cambridge. The rare *Ilyobates nigricollis* is recorded from Wicken Fen, and the writer has taken it under cut grass refuse at Chippenham Fen. The beautiful *Myrmedonia collaris* occurs in Wicken Fen; Messrs Chitty, Bouskell and Donisthorpe took it in some numbers in company with the ant *Myrmica lævinodis* in cut sedge refuse in 1900. The pretty little *Hygronoma dimidiata* is not uncommon in stems of reeds in Wicken Fen. The little *Mycetoporus longicornis*, always a rare insect, has been taken by the writer under vegetable refuse in Wicken Fen. Several specimens of the beautiful *Staphylinus fulvipes* were taken by Messrs Champion, Gorham and Walker under cut grass at Chippenham Fen in 1898; the rare *Ocypus fuscatus* has occurred to the writer at Upware under a stone. *Xantholinus tricolor* occurs under vegetable refuse in Chippenham Fen and Wicken Fen, and the rarer *X. glaber* was taken by Dr Power at Grantchester. In the genus *Stenus*, *S. proditor*, chiefly a fen species, may be mentioned from Wicken. The little *Thinobius brevipennis*, purely a fen insect, has been taken at Quy Fen.

Passing on to the large order *Clavicornia*, in the genus *Anisotoma*, Wicken and Burwell Fens are two of the very few localities recorded for the rare *A. obesa*; and *A. ovalis* is not uncommon by evening sweeping in the former locality. In the carrion-feeders the writer has taken the large *Necrodes littoralis* under fish refuse at Upware. *Silpha tristis* occurs at roots of grass at Wicken Fen, the rare *S. nigrita* was taken by Power on the Gogmagog Hills, and the spotted *S. 4-punctata* has been recorded from Gamlingay on oaks.

Among the very small *Tricopterygidæ*, *Trichopteryx championis* appears only to have occurred at Wicken Fen, and the very rare *Ptilium affine* and *P. incognitum* seem to be confined to the same locality. In the "ladybirds," *Coccinella hieroglyphica* may be mentioned, both the type and the black form occurring in Wicken Fen: the very local *Murmidius ovalis* was taken by Dr Power at Madingley Wood. The extremely scarce *Nemosoma elongatum* was taken by Professor Babington near Cambridge in May, 1834. The pretty little spotted *Psammœchus bipunctatus* is not uncommon in cut sedge and stems of reeds in the fens. *Antherophagus nigricornis*, which is parasitic on certain bees, may be swept not uncommonly off flowers in Wicken Fen.

The very rare *Cryptophagus schmidti* has only been taken at Wicken Fen (by Mr Champion), and at Whittlesea Mere (by Mr E. W. Janson). Of the *Parnidæ* the writer took one of the few British specimens of *Parnus nitidulus* in a ditch at Chippenham Fen; it has since been taken in the same locality by Mr Bouskell.

The most interesting record of the *Lamellicornia* is *Copris lunaris*, which, according to the Rev. L. Jenyns, occurred in great plenty in a field near Melbourne in the spring of 1828, but apparently has not been noticed since.

Of the *Serricornia* the little *Cryptohypnus quadripustulatus* may be swept in Wicken and Burwell Fens, and *Corymbites tesselatus* is not uncommon in the former locality. The last specimen that has occurred in Britain of the very rare *Ludius ferrugineus* was taken by a schoolboy, on a poplar on the river Cam, between Cambridge and Grantchester; it has also been recorded from Bottisham. *Platycis minutus* was taken in some numbers by Messrs Champion, Gorham and Walker, in Chippenham Fen in 1898, out of old birch stumps. The "Glowworm," *Lampyris noctiluca*, is not uncommon in the fens. *Silis ruficollis* is numerous in some years only, and then may be taken by sweeping in Wicken and Burwell Fens. *Anthocomus rufus*, a purely fen insect, is plentiful in Wicken Fen on the meadow-sweet in August. The very rare *Anthocomus terminatus* sometimes occurs; Messrs Beare, Bouskell and Donisthorpe took a small series in Wicken Fen in 1896, and Mr Morley records it from Chippenham Fen.

There are several interesting Longicornes, which are found in the fens. *Aromia moschata*, the well-known "Musk Beetle," is found about willows at Wicken, and a few years back it was common at Upware. The large *Saperda carcharias* is plentiful on poplars at Wicken, and also used to be frequent at Upware. *Agapanthia lineatocollis* is often abundant on thistles at Chippenham Fen, and is, in some years, not uncommon at Wicken, where a melanic form occurs at times. The beautiful *Oberea oculata*, one of the prizes of the coleopterist, appears to be confined to Wicken Fen in this country, where it may be found in August on the sallow bushes; it was plentiful in 1898.

Coming to the *Chrysomelidæ* the rare *Donacia dentata* is sometimes plentiful on water plants in Wicken Fen in August, and *Donacia sparganii* is also to be found in the same locality. The only British specimen of *Cryptocephalus primarius* was taken by Dr Power many years ago on the Gogmagog Hills. The beautiful *Chrysomela graminis*, chiefly a fen species, occurs in plenty on water-mint in Wicken Fen, and is recorded from Soham, Ely and Burwell Fens. *Melasoma populi* is plentiful on young poplars in Wicken Fen, occurring in August in all its stages. The extremely rare *Adimonia œlandica* was taken in numbers in Wicken Fen by Mr Blatch in August 1878. The same fen is one of the few recorded localities for the rare *Longitarsus waterhousei*, and *Phyllotreta sinuata* is to be found there, and at Quy Fen.

Passing now to the *Heteromera* the coast species *Cteniopus sulphureus* is common on flowers in Wicken Fen; Cambridgeshire is one of the few counties where the "Blister Beetle," *Lytta vesicatoria*, occurs; it has been abundant during the last few years at Newmarket, on privet hedges and ash. In 1901 the writer ran it down in its old locality near the Gogmagog Hills on ash trees.

Finally we come to the *Rhynchophora*, in which *Lixus paraplecticus*, a curious weevil, with its elytra terminating in two long points, is one of the most interesting beetles of the Fens; it is said to have disappeared before drainage, but this is not the case so far as Wicken Fen is concerned, where it occurs in abundance on *Sium latifolium*, the larvae living in the stems of this plant. The perfect insect is covered with a yellow pollen-like dust, which it has the power to renew during life. The pretty little *Nanophyes lythri* is abundant on *Lythrum* in Wicken Fen, and the local and rare *Dorytomus salicinus* may be taken in plenty by beating the sallow bushes in Wicken Fen. *Hylesinus crenatus* and *H. oleiperda*, both occurring in ash, are to be found at Chippenham; the latter bores into the small twigs at the end of the branches, from which it may be beaten.

THE

LEPIDOPTERA OF CAMBRIDGESHIRE.

By Wm Farren.

RHOPALOCERA.

ALTHOUGH Cambridgeshire cannot be considered a good butterfly county, yet with records made in the earlier part of last century, most of which are quite reliable, a list might be formed comprising some sixty species, which is within half-a-dozen of the complete British list; at the present time however it would not be easy to find more than thirty-five species.

Most important of these, the Swallow-tailed butterfly, *Papilio machaon*, L., still exists in considerable numbers in Wicken Fen, but it is no longer found in the surrounding fens and north of Ely, where it was plentiful fifty or sixty years ago. *Aporia crataegi*, L., probably, and *Leucophasia sinapsis*, L. certainly occurred in the woods near Gamlingay in the time of Jenyns, the latter not uncommonly; it was taken there by Prof. Babington as late as 1835. *Pieris daplidice*, L., has been frequently taken, but not recently, on the chalk between Cambridge and Newmarket. In years notable for a visitation of "Clouded yellows," the two species of *Colias*, *C. edusa*, Fb., and *C. hyale*, L., have been taken in most parts of the county, but nowhere in such abundance as on the chalk in the south-east. There are records for all the "Fritillaries," of which the two small species of *Argynnis*, *A. selene*, Schiff., and *A. euphrosyne*, L., may still occur at

Gamlingay; *A. aglia*, L. was common in Bottisham and the
adjacent fens, up to thirty years ago, and *Melitaea aurinia*,
Rott. also occurred in various parts of the fens, but has not
been recently recorded. Of the genus *Vanessa*, *V. C-album*,
L. was not uncommon at Shelford about 1840, and *V. antiopa*,
L. has been frequently taken in the county; the remaining
species are all more or less common.

Thecla w-album, Knoch. occurs at Madingley and other
localities in the west. Mr P. T. Gardner finds it plentiful,
sometimes abundant, at Conington; it is doubtful whether
any other species of this genus now occurs in the county.
Polyommatus dispar, Haw. had probably ceased to exist in
Cambridgeshire long before it finally disappeared from the
Huntingdonshire Fens, but Mr Wagstaff took a specimen on
Bottisham Fen in 1851, the year Whittlesea Mere was
drained.

Lycaena acis, Schiff., now extinct, occurred plentifully
around Cambridge up to fifty years ago; *L. argiolus*, L.
which was hardly known in the county previous to 1900,
turned up then in several localities near Cambridge, and it
has occurred regularly since, not only in and near the town,
but at Shelford and the Gogmagogs, and in the extreme
west of the county. The common *Hesperidae*, including
Hesperia comma, L., are all plentiful; *Hesperia lineola*, only
noticed as occurring in Britain ten or twelve years ago, is
common in Burwell and the adjacent Fens, also in the
extreme west, and in the south.

HETEROCERA.

Sphinges.

With the HETEROCERA it becomes necessary, in order not
to exceed the space allotted to this article, to refer only to
species of exceptional local interest.

Among the *Sphingidae*, all of which are plentiful, with
the exception of the occasional migrants, *Acherontia atropos*,

L., and *Sphinx convolvuli*, L., may be mentioned; larvae and pupae of the former are found plentifully in some seasons in the large Fen potato fields. *Deilephila galii*, Schiff. was fairly common on the chalk in 1888; *Macroglossa bombyliformis*, Och. used to occur on Horningsea Fen and more recently at Chippenham.

Of the *Sesiidae, Trochilium apiformis*, Clerck, is common in the poplars which surround and shelter the Fen farms; *Sesia formicaeformis*, Esp., although much more restricted in its range than formerly, still occurs in some osier beds near the Ouse.

Bombyces.

Fully two-thirds of the BOMBYCES might be claimed for the county, but comparatively few are of great local interest. *Nudaria senex*, Hb., a genuine bog insect, is abundant at Wicken, and also occurs on the marshy banks of the upper Cam; *Nemeophila russula*, L. occurred in the time of Jenyns in Horningsea Fen and has been taken at Chippenham; *Spilosoma urticae*, Esp., plentiful enough in the Fens up to ten years ago is gradually becoming scarce. *Callimorpha dominula*, L. occurs rarely at Chippenham; it was abundant in Wicken and the neighbouring Fens up to the seventies of last century, when it rapidly became scarce and disappeared about the same time that *Laelia coenosa*, Hb. became extinct; the latter species occurred in such profusion in the mid fifties that some other explanation than its being collected in large numbers seems necessary to account for its somewhat sudden disappearance twenty years later. *Macrogaster castanae*, Hb. is rather more plentiful at Chippenham than at Wicken, but it is not by any means scarce in the latter locality: recently taken specimens are considerably smaller than those which occurred at Whittlesea Mere, the size depending apparently on the size of the reed in which they feed. *Ocneria dispar*, L., now probably extinct as a British insect, occurred sparingly in the Fens in the time of Jenyns, as also did *Orgyia gono-*

stigma, Fb. *Dasychira fascilina*, L. is plentiful in some seasons on the Hills Road, and *Lasiocampa quercifolia*, L. is common in the Fens.

All of the *Drepanulidae*, except *D. sicula*, Hb., may be found in the county. *Stauropus fagi*, L. is occasionally found on the Gogmagogs, where Mr Jones has taken a specimen of *Notodonta dictaeoides*, Esp.

Of the *Cymatophoridae*, *C. octogesima*, Hb. is most worthy of notice; it is to be found in fair numbers in most parts of the county; more especially in the Fens.

Noctuae.

Among the NOCTUAE, the *Leucaniidae*, reed and grass feeding insects, are naturally confined to marshy places, and many of them are only to be found in the Cambridgeshire Fens. Before enumerating these however, mention must be made of the Cambridge form of *Bryophila muralis*, Forst., known as *B. impar*, Warren; *B. muralis*, Forst., is a coast species, and the isolated inland race at Cambridge has acquired a peculiar facies which distinguishes it from *B. muralis*, Forst., in all of its British localities. It occurs sparingly on the old walls of the town; many of the best walls have been recently demolished, but *impar* is so widely distributed in the town that it is not likely to be exterminated. Less than twenty years ago *Acronycta strigosa*, Fb., could be found all along the chalk on the edge of the Fens, from the Gogmagogs to Fordham, but it has become very scarce during the last twelve years and is now almost extinct. *Arsilonche albovenosa*, Göze, may still be found in the Fens, but is not nearly so plentiful as it was fifteen years ago. *Leucania obsoleta*, Hb. occurs sometimes in fair numbers near Ely, where *Senta maritima*, Tausch. may also be found, but in smaller numbers; *Leucania straminea*, Tr., is not common, but occurs in the reed beds of the upper Cam, where *Nonagria neurica*, Hb. was not uncommon a few years ago, but its favourite reed beds have become dry and overgrown with rank herbage, so that latterly it has become somewhat scarce. *Calamia phragmitidis*, Hb., is common. *Meliana flammea*, Curt., is plentiful at Wicken and Chippenham, as also are *Coenobia rufa*, Haw., *Tapinostola fulva*, Hb., and *T. hellmanni*, Evers.; the latter has been taken at Shelford and Boxworth; at the latter locality a specimen of *Nonagria geminipuncta*, Hatch., was captured by Mr P. T. Gardner. *N. arundinis*, Fb., and *N. lutosa*, Hb., are both common throughout the Fens. The most noteworthy of the *Apameidae* are *Luperina cespitis*, Fb., occasionally taken in the west, *Mamestra abjecta*, Hb., which is very scarce at Wicken, and *M. albicolon*, Hb., in the extreme east of the county; *Apamea gemina*, Hb., *A. unanimis*, Tr., and *A. leucostigma*, Hb., are common in the Fens, and *A. ophiogramma*, Esp., is not scarce on the banks of the upper Cam. *Miana literosa*, Haw., occurs sparingly, and *arcuosa*, Haw., plentifully in many parts of the county. The *Caradrinidae* are well represented; most of the British taken specimens of the rare *Hydrilla palustris*, Hb., have been captured at "light" in the fens, a few were taken in 1876 or 1877, not more than a dozen odd specimens turned up between then and 1897, when about half-a-dozen were taken, and in 1898 the captures numbered about 50; the writer knows of only one specimen which has been taken since. *Agrotis obscura*, Brahm., occurs at Wicken and other parts of the county. *Noctua stigmatica*, Hb., has become more generally plentiful in the last ten years, it is common in some seasons in the west of the county. Most of the species of this genus may be found in the county, and all of *Triphaena*; a specimen of *T. orbona*, Hufn., = *subsequa*, Hb., having been taken by Mr Alfred Jones at Chippenham in 1892 and two by Mr Thornhill at Boxworth in 1903.

Of the *Orthosiidae* and *Cosmiidae*, most of which occur, mention may be made of *Cirrhoedia xerampelina*, Hb., which is not uncommon near Cambridge and towards the western boundary, and *Calymnia pyralina*, View., which Mr E. H. Thornhill finds not uncommon in some seasons at Boxworth; it has also been taken at Wicken.

Among the *Hadenidae, Aplecta advena,* Fb. is not rare, and *A. occulta,* L. has been taken in the west.

The extremely rare *Hadena satura,* Hb., was taken at Wicken by Mr Tutt in 1893. *H. atriplicis,* L., abundant in the Fens thirty years ago is now almost extinct.

The only species which call for mention in the *Xylinidae* are *Xylina semibrunnea,* Haw., and *Asteroscopus sphinx,* Hufn., which are sparsely distributed throughout the greater part of the county. Of the *Plusiidae, Plusia chryson,* Esp., occurs in Chippenham Fen, *P. festucae,* L., at Wicken and on the banks of the rivers; *P. moneta,* Fb., which was taken in Cambridge in 1901, the year after it was first discovered in England, has been steadily increasing in numbers since, and may now be considered common and well established, and a specimen of *P. interrogationis,* L., was taken in the grounds of Jesus College in 1886. Of the remaining genera of *Noctuae, Acontia luctuosa,* Esp. is sometimes plentiful in the west, *Erastia fasciana,* L. occurs sparingly, and *Bankia argentula,* Hb. abounds at Chippenham, where *Hydrelia uncula,* Clerck, also occurs, but not so plentifully as at Wicken. *Aventia flexula,* Schiff., is taken by Mr Thornhill at Boxworth, and Mr Gardner at Conington. *Herminia cribralis,* Hb. is common, and odd specimens turn up of *Hypenodes albistrigalis,* Haw., and *H. costaestrigalis,* St., in the Fens.

Geometrae.

Although the GEOMETRAE are numerously represented, comparatively few are of great local interest: the black var. *doubledayaria,* Mill., of *Amphidasys betularia,* L., has been occasionally taken in and near Cambridge and Ely. The beautiful little *Hyria muricata,* Hufn. is common in Wicken Fen and may be seen flying in abundance at sunrise. *Geometra vernaria,* Hb., *Melanippe procellata,* Fb., and *Eupethecia isogrammata,* H.-S., are common at Cherryhinton and other places where *Clematis vitalba* grows; Mr Jones once bred *Eupethecia consignata,* Bork. from mixed larvae beaten from hawthorn on the Gogs. *E. plumbeolata*, Haw., *E. pygmaeata*, and *E. valerianata*, Hb., are not uncommon at Wicken, and *E. subciliata*, Gn. occurs near Cambridge, but more commonly at Boxworth. *Collix sparsata*, Hb., and *Lobophora sexalisata*, Hb., are plentiful at Wicken. *Anticlea cucullata*, Hufn. occurs all over the chalk, and *A. berberata*, Schiff., although having its head-quarters at Bury St Edmund's, in Suffolk, has been taken by the writer on the Gogmagogs. *Coremia quadrifasciata*, Clerck, occurs sparingly in the west, and also near Cambridge. *Phibalapteryx polygrammata*, Bork., now extinct, was not uncommon in Burwell Fen, but repeated efforts to rediscover it have met with failure.

Two unexpected species are recorded by Jenyns, *Eucosmia undulata*, L.;. and *Cidaria psitticata*, Schiff., the former near Cambridge, and the latter at Bottisham.

Cidaria sagittata, Fb., which may be found commonly all through the Fens from Bottisham to Chatteris, almost disappeared about 1892, to turn up as plentiful as ever six years after. *Mesotype virgata*, Rott., occurs in the Devil's Ditch, and Jenyns records *Tanagra atrata*, L., as being common on Gamlingay Heath in the spring of 1829.

Pyralides.

A good number of the Pyralides may be collected in Cambridgeshire; *Aglossa cuprealis*, Hb., may sometimes be found commonly in outhouses, especially in the Fen villages. *Scoparia pallida*, St. is abundant in Wicken and Chippenham Fens. *Spilodes sticticalis*, L., and *S. palealis*, Schiff., have both been taken. *Orobena extimalis*, Scop., and *Perinephele lancialis*, Schiff., although very rare have been taken by the writer, the former at Fulbourn, and on the Suffolk boundary, the latter at Chippenham. The beautiful *Nascia cilialis*, Hb. occurring nowhere else in Britain comes to "light" abundantly in Wicken Fen. *Acentropus niveus*, Olis. is sometimes common on the river near Ely.

Pterophori.

Of the *Pterophori* mention may be made of *Oxyptilus laetus*, Zell., which is found on the Suffolk border, *O. parvidactylus*, Haw., taken by the writer in good numbers on the Gogmagogs, and *O. pilosellae*, Zell., taken on the Devil's Ditch by Mr F. Bond, who also found *Oedematophorus lithodactylus*, Tr. commonly at Whittlesford. *Leioptilus microdactylus*, Hb., is common in Wicken and Chippenham Fens, and at the latter locality *Aciptilia galactodactylus*, Hb. also occurs; *A. tetradactyla*, L., has been taken by the writer at Fulbourn.

Crambi.

Among the *Crambidae*, *Chilo phragmitellus*, Hb. is abundant, and *Schoenobius mucronellus*, Schiff., and *S. gigantellus*, Schiff., are rare at Wicken.

Of the rarer species of *Crambus*, *sylvellus*, Hb., *C. uliginosellus*, Zell., *C. hamellus*, L., and *C. contaminellus*, Hb., are recorded in the "Fenland" list; the first two at Wicken, the others at Wisbech. The writer has found *C. falsellus*, Schiff. at Burwell, and *C. selasellus*, Hb. is plentiful in parts of the Fens.

Of the *Phycidae*, mention may be made of *Euzophera pinguis*, Haw., which occurs locally in various parts of the Fens, and notably on a large ash tree in Lensfield Road, Cambridge. *Rhodophoea advenella*, Zinck. is fairly common in some hawthorn hedges, and *Oncocera ahenella*, Zinck. occasionally turns up on the Fleam Dyke.

Tortrices.

Although among the Tortrices there are many local and scarce species to be found in Cambridgeshire, comparatively few are exclusively Fenland species. *Tortrix dumetana*, Tr., is extremely local, but abundant at Wicken, as also are *Peronea shepherdana*, St., and *P. hastiana*, L., of which latter some very beautiful varieties occur; *P. comparana*, Hb., and *P. schalleriana*, L., are generally common: the writer has taken a peculiar small form of the latter, mostly of the var. *latifasciana*, Haw., flying in June among meadow-sweet. *Rhacodia caudana*, Fb. occurs locally. *Dictyopteryx lorquiniana*, Dup., one of the most purely Fen Tortrices, is common at Wicken and Ely. *Penthina fuligana*, Hb. used to be found commonly in Wicken Fen, but has become decidedly rare in the last ten years; *P. sellana*, Hb. may be found at Wicken, but is scarce, as also is *Hedya servillana*, Dup., which the writer has also found at Fulbourn. *Antithesia salicella*, L. is not uncommon, but local. *Sericoris fuligana*, Haw., is sometimes plentiful in Wicken Fen. *Orthotaenia antiquana*, Hb., is common, but *Phtheochroa rugosana*, Hb., although widely distributed in the county is scarce. *Clepsis rusticana*, Tr. occurs, but rarely, at Wicken and Chippenham. Several interesting species of *Phoxopteryx* are to be found in Wicken and other Fens, among which may be mentioned *P. siculana*, Hb., *P. unguicella*, L., *P. inornatana*, H.-S., *P. diminutana*, Haw., and *P. paludana*, Bar.; *Paedisca rufimitrana*, H.-S. was taken by Mr Jenkinson at Cambridge, and *P. solandriana*, L., and *semifuscana*, St., occur at Wicken; *P. oppressana*, Tr., is not rare but very local in the Fens. The writer has taken *Coccyx subsequana*, Haw. at Chippenham.

Opadia funebrana, Tr. is more plentiful in Cambridgeshire than elsewhere, but specimens are difficult to procure, although the larvae are but too plentiful in plums. *Stigmonata orobana*, Tr. is not uncommon at Wicken. *Dicrorampha alpinana*, Tr. may be found in gardens in Cambridge.

Catoptria expallidana, Haw., occurs in small numbers in Wicken Fen. *Choreutes myllerana*, Fb. is common on the banks of the upper Cam, and *Symaethis pariana*, L. occurs sparingly near Cambridge. An interesting form of *Eupoecilia vectisana*, Westw., a species otherwise confined to a few seacoast localities, occurs in Wicken Fen, where *E. notulana*, Curt., *E. griseana*, Haw., and *E. rupicola*, Curt., may also be found, and specimens of *E. degreyana*, MacLach. have been recorded at the same locality. Of the remaining Tortrices the most noteworthy is *Argyrolepia schreberiana*, Fröl., which was plentiful near Ely station fifteen years ago, but has not been found for some years.

Tineae.

Among this, the last and most numerous family of Lepidoptera, we find a considerable number of species, especially among the *Gelechiidae*, which are exclusively Fen-dwellers. *Dasystoma salicella*, Hb., which generally occurs sparingly in hawthorn hedges, the writer once saw flying in large numbers on Hills Road. *Tinea pallescentella*, Sta., at one time considered a scarce insect, is very plentiful in some Cambridge houses. *Micropteryx calthella*, L. is common throughout the Fens, and *M. thunbergella*, Fb., at Chippenham. *Nemophora metaxella*, Hb., a very local species, is plentiful in the Fens.

The beautiful species of *Adelidae*, *Adela croesella*, Scop., and *A. degeerella*, L., occur at Chippenham, *Nematois cupriacellus*, Hb., occasionally at Fulbourn and other localities, and *N. fasciellus*, Fb., sometimes in fair numbers at Burwell. The two local and pretty species *Anesychia funerella*, Fb., and *A. decemguttella*, Hb., are found commonly in the Fens, the former in Wicken and surrounding Fens, the latter at Chippenham. The genus *Depressaria* is well represented, especially in the Fens; the most noteworthy species are *D. costosa*, Haw., *D. flavella*, Hb., *D. pallorella*, Zell., *D. umbellana*, St., the *var. rhodocrella*, H.-S. of *D. subpropinquella*, Sta., *D. purpurea*, Haw., *D. liturella*, Hb., *D. conterminella*, Zel., *D. angelicella*, Hb., *D. yeatiana*, Fb., *D. ciliella*, Sta., and *D. chaerophylli*, Zell. Among the remaining genera of the *Gelechiidae* are to be found the greatest number of purely fen species of all the Lepidoptera with the exception perhaps of the *Leucaniidae*. Of these mention may be made of *Gelechia muscosella*, Zell., *G. divisella*, Doug., and *G. rhombella*, Schiff.; *Brachmia lathyrella*, Sta.; *Bryotropha umbrosella*, Zell., and *B. basaltinella*, Zell.; *Lita acuminatella*, Sircom., *L. fraternella*, Dougl., *Nannodia stipella*, Hb., *Ergatis subdecurtella*, Sta.; *Doryphora palustrella,* Dougl., *D. lutulentella,* Zell., *D. oblitella,* Dbl., *D. morosa,* Mühlig, and *D. quaestionella,* H.-S.; *Ceratophora rufescens,* Haw., and *C. inornatella,* Dougl., and *Cladodes gerronella,* Zell., most of which may be found in Wicken Fen, where also *Parasia lappella,* L., and *P. metzneriella,* Sta., occur. *Oecophora minutella,* L., and *Oe. unitella,* Hb., are with others beaten from thatch at Wicken; *Oe. fulviguttella,* Zell., is common in the Fen. *Cataplectica farreni,* Walsm., discovered, as new to science, by the writer in 1892, is generally abundant in Cherryhinton chalk-pit, and occurs also on other parts of the chalk among its food plant *Pastinaca sativa. Oecogenia quadripunctata,* Haw., was taken in Cambridge by Mr F. Bond ("Fenland" list). *Butalis chenopodiella,* Hb., and *Acrolepia granitella,* Tr., have both been taken near Cambridge, but the writer has failed to find them; he has found *A. pygmaeana,* Haw., however in some abundance. *Glyphipteryx thrasonella,* Scop., *G. cladiella,* Sta., and *G. forsterella,* Fb., occur at Wicken. *Aechmia dentella,* Zell., is sometimes common near Cambridge. *Gracillaria imperialella,* Mann. is sometimes plentiful at Wicken, where *G. auroguttella,* St., has also been taken, and the writer has found *G. omissella,* Dougl., abundant near Linton. *Ornix avellanella,* Sta., and *O. guttea,* Haw., occur at Cambridge, and *O. fagivora* among the beeches on the Gogmagogs. Of the many species of *Coleophora* may be mentioned *C. fabriciella,* Vill., *C. melilotella,* Scott, *C. troglodytella,* Dup., *C. apicella,* Sta., and *C. palliatella,* Zinc., occurring in the Fens, and *C. alcyonipennella,* Kol., *C. siccifolia,* Sta., *C. lixella,* Zell., and *C. laricella,* Hb. near Cambridge.

In 1892 the writer found the local and beautiful *Stathmopoda pedella,* L., in fair numbers in Chippenham Fen, but the alder bushes it frequented were destroyed and only odd specimens have been taken since. *Cosmopteryx lienigiella,* Zell. is fairly common in Wicken and Chippenham Fens, and *C. orichalcea,* Zell., less common, and only in Wicken Fen. *Chauliodus illigerellus,* Hb. occurs plentifully in the Fens, and *C. chaerophyllellus,* Göze, is sometimes abundant, and in great variety of colour on the Hills Road and at Cherryhinton, feeding principally on *Pastinaca sativa*.

Laverna decorella, St. occurs rarely at Cambridge, *L. atra*, Haw., is sometimes abundant; *L. phragmitella*, Bent. was formerly common in the Fens, and *L. lacteella*, St. has been taken at Wicken by Lord Walsingham ("Fenland" list). There is a record in the same list of *Chrysoclista schrankella*, Hb., taken at Cambridge by Mr Warren. Among the many rare and local species of *Elachista* occurring in the Fens the writer has taken *E. gleichenella*, Sta. at Chippenham, *E. paludum*, Frey., in a bog near the Suffolk border, and *E. pollinariella*, Zell., in fair numbers on the Gogmagog Hills.

Many species of *Lithocolletis* occur, but none of great local interest, as they belong generally to the higher land and the woods; most of the oak-feeding species may be found near Madingley. *Phyllocnistis suffusella*, Zell., and *P. saligna*, Zell., both occur in plenty, the latter however is very local. *Opostega crepusculella*, Fisch. is common at Wicken, where the scarce *O. auritella*, Hb. has also been taken. The same remarks apply to *Nepticula* as to *Lithocolletis*, the oak-feeding species being particularly plentiful in one plantation near Madingley—of these *N. basiguttella*, Hein. is rare and it is difficult to obtain specimens, but *N. quinquella*, Bedell, a very local species, is in some seasons most abundant; in November 1891 hardly an oak-leaf was free from the mines of the larvae; the writer counted 73 in one leaf! A visit was paid to the wood in the following May, when the brilliant little moths were to be seen all over the trunks of the oaks, the metallic spots glistening in the bright sunshine, giving to the trees quite a silver-spangled appearance.

THE DIPTERA OF CAMBRIDGESHIRE.

By Jas. E. Collin.

The Diptera are for many reasons by far the most difficult Order to deal with in making a complete account of the characteristic fauna of any particular county, and Cambridgeshire is no exception to the rule.

The number of collectors of Diptera in England is very small, and those who study them may almost be reckoned on the fingers of one's hands, while the number of those who have correctly named their captures and published any Lists is smaller still; consequently our knowledge of the distribution of species in England is very slight, and a species may have been considered local, or rare, merely because it has never been collected, or if collected has not been named and recorded, from any other locality. It is very probable that the species of Diptera are more widely distributed in England than any other Order and that it is only a question of looking for them in the right places.

In this account of the Cambridgeshire Diptera it is proposed to divide the county into two districts, viz.:

1. The Northern or Fen District; including practically all that part of the county north of Newmarket and Cambridge.

2. The Southern District.

By this means we make it easier to refer to those Diptera occurring in the Fens, among which there is greater probability of species being confined to Cambridgeshire only, or occurring in numbers in that county and rare elsewhere, because in no other English county is there such another piece of original undrained fenland as is found at Wicken.

A history of this Fen District (which is far the most important from an entomological point of view) will be found elsewhere, and it is only necessary to state here that the greater part has been drained and converted into exceedingly rich agricultural land, by confining the flow of water to certain artificial channels, and pumping the water from the surrounding land into those channels; only about 300 acres remain to give anyone an idea of what the Fen District was like before it was drained. These 300 acres, which are known as "Wicken Fen," possess a fauna and flora of remarkable interest, and it is to be hoped that they may never suffer the fate that has befallen the rest of the Fen-land. At Chippenham there is another piece of fen-land, which was probably close to the edge of the original Fen District, where it merged into the sandy heaths beginning at Newmarket and culminating in the wild arid country round Barton Mills and Elvedon (Suffolk) and Thetford (Norfolk). This piece of fen-land is also exceedingly rich entomologically, and possesses in many respects a fauna and flora distinct from Wicken Fen because it is more or less surrounded by woods, and consequently in addition to the usual marsh-loving insects and plants, one finds many species peculiar to damp woods; it is the only locality in England where *Teuchophorus simplex*, Mik., has been found.

The geological and topographical features of the Southern District show much greater variation; there is the valley of the Cam and those of its tributaries in the South-West; a range of chalk hills, which rise to a height of from 300 to 400 feet and include the Gog Magog Hills near Cambridge, enter the county at Royston and extend from south to south-east viâ Linton to Newmarket; south of these hills the country is extensively wooded, and such large woods as Woodditton Park and The Wydghams would probably well repay any collector who would undertake to work them, for such comparatively rare species as *Volucella inflata*, F., *Criorrhina asilica*, Fln.,

Helophilus frutetorum, F., and *Xylota abiens*, W. were once taken in an odd half-hour's collecting, at a hedge of whitethorn in blossom, under Woodditton Wood.

Before proceeding to enumerate the Diptera of the Fen District some mention must be made of the Rev. E. L. Jenyns of Swaffham Prior, who collected all Orders of Insects seventy years ago in the immediate neighbourhood of his home. An examination of the Diptera collected by him shows that he took two or three species that have not been recorded from the county since, such as *Tabanus bovinus*, L., our largest British Dipteron, which he says was not uncommon at Ely and Bottisham, *Atylotus rusticus*, F. which he found at Cambridge, and *Thyreophora furcata*, F. taken at Ely; he also took *Orthoneura brevicornis*, Lw., *Liogaster metallina*, F., *Chilosia honesta*, Rnd., *Pyrophaena granditarsa*, Forst., *Catabomba selenitica*, Mg., *Xanthogramma citrofasciatum*, De G., *Sphegina clunipes*, Fln., *Eristalis aeneus*, Scop., *Gastrophilus haemorrhoidalis*, L., and *Stenopteryx hirundinis*, L.

Specimens of *Cecidomyidae*, *Mycetophilidae*, and *Chironomidae* have been but little collected in the Fen District, though one cannot help thinking that it would prove a good locality for these families, especially Chippenham Fen for *Mycetophilidae*.

Among the *Bibionidae*: *Scatopse bifilata*, Hal. and *S. coxendix*, Verr., *Bibio leucopterus*, Mg. (in swarms), *B. venosus*, Mg., and *B. lepidus*, Lw. have occurred at Chippenham, and *Scatopse geniculata*, Ztt., and *S. inermis*, Ruthé, at Wicken.

Of the *Limnobidae*: *Limnobia analis*, Mg., *Acyphona maculata*, Mg., *Molophilus ochraceus*, Mg., *Rhypholophus varius* Mg., *Erioptera flavescens*, Mg., *Limnophila nemoralis*, Mg., and *Pedicia rivosa*, L., may be mentioned; while in the *Tipulidae* the following have been captured: *Tipula nigra*, L., *T. longicornis*, Schum, *T. scripta*, Mg., *T. luteipennis*, Mg., and *T. lunata*, L.

Among the *Stratiomyidae* the two large handsome species of *Stratiomys*, viz. *S. chamaeleon*, L., and *S. potamida*, Mg. occur sparingly at Chippenham, and *S. furcata*, F. has been taken; *Oxycera morrisii*, Curt. and *O. trilineata*, F., *Nemotelus nigrinus*, Fln., and the commoner species *N. pantherinus*, L., and *N. uliginosus*, L. all occur. *Microchrysa flavicornis*, Mg., *Beris vallata*, Forst., and *Xylomyia marginata*, Mg. have been taken at Upware close to Wicken.

All the commoner species of *Tabanidae* occur in the district, but there is nothing else of note to mention until we come to the *Empidae*, among which we find *Cyrtoma melaena*, Hal., *Rhamphomyia spissirostris*, Fln., *R. dissimilis*, Ztt., *R. atrata*, Mg., and *R. umbripennis*, Mg., *Empis livida*, L., *E. trigramma*, Mg., *E. punctata*, Mg., *E. caudatula*, Lw., and *E. aestiva*, Lw., *Trichina elongata*, Hal., *Hemerodromia precatoria*, Fln., *Ardoptera irrorata*, Fln., and *A. guttata*, Hal., *Lepidomyia melanocephala*, F., *Drapetis nervosa*, Lw., and *D. pusilla*, Lw., *Tachypeza fuscipennis*, Fln., and *Chersodromia speculifera*, Walk., a little known species which however is not uncommon in Chippenham Fen in heaps of cut sedge in the winter and spring.

In the next family *Dolichopodidae* we find a whole host of species occurring in the Fen District, many of which had not been recorded as British, or had been considered rare, until collected there; among them may be mentioned: *Psilopus longulus*, Fln., and *P. contristans*, W., *Eutarsus aulicus*, Mg., *Dolichopus nubilus*, Mg., *D. pennatus*, Mg., *D. urbanus*, Mg., *D. longicornis*, Stann., *D. puncticornis*, Ztt., *D. linearis*, Mg., *D. griseipennis*, Stann., *D. simplex*, Mg., *D. festivus*, Hal., *D. trivialis*, Hal., and *D. brevipennis*, Mg., *Poecilobothrus nobilitatus*, L., *Hercostomus chrysozygos*, W., *H. gracilis*, Stann., *H. nigriplantis*, Stann., *H. plagiatus*, Lw., and *H. nanus*, Mcq., *Hypophyllus discipes*, Ahr., and *H. obscurellus*, Fln., *Gymnopternus metallicus*, Stann., and *G. assimilis*, Staeg., *Chrysotus pulchellus*, Kow., and *C. microcerus*, Kow., *Argyra diaphana*, F., *A. confinis*, Ztt., *A. argentina*, Mg., and *A. leucocephala*, Mg., *Leucostola vestita*, W., *Syntormon pumilus*, Mg., and *S. pallipes*, F., *Machaerium maritimae*, Hal., *Xiphandrium appendiculatum*, Ztt., and *X. caliginosum*, Mg., *Medeterus diadema*, L., *M. flavipes*, Mg., *M. jaculus*, Mg., and *M. truncorum*,

Mg., *Hydrophorus balticus*, Mg., and *H. bisetus*, Lw., *Campsicnemus curvipes*, Fln., and *C. picticornis*, Ztt., *Teuchophorus spinigerellus*, Ztt., and the rare *T. simplex*, Mik., *Lamprochromus elegans*, Mg., *Bathycranium bicolorellus*, Ztt., *Micromorphus albipes*, Ztt.

In the *Pipunculidae* we find *Pipunculus furcatus*, Egg., *P. zonatus*, Ztt., *P. modestus*, Hal., *P. unicolor*, Ztt., *P. varipes*, Mg., *P. stroblii*, Verr., *P. confusus*, Verr., *P. haemorrhoidalis*, Ztt., *P. xanthopus*, Thoms., and *P. maculatus*, Walk.

The most noteworthy captures among the *Syrphidae* include *Chilosia nebulosa*, Verr., common at Chippenham in the spring, *C. praecox*, Ztt., *C. vernalis*, Fln., *C. proxima*, Ztt., and *C. cynocephala*, Lw., *Platychirus scambus*, Staeg., *P. fulviventris*, Mcq., and *P. angustatus*, Ztt., *Syrphus tricinctus*, Fln., *S. labiatarum*, Verr., and *S. barbifrons*, Fln., *Ascia geniculata*, Mg., *Brachyopa bicolor*, Fln., *Rhingia campestris*, Mg., *Helophilus trivittatus*, F., *H. hybridus*, Lw., *H. versicolor*, F., and *H. lineatus*, F., *Chrysochlamys cuprea*, Scop., and *Chrysotoxum festivum*, L., and *C. bicinctum*, L.

Our knowledge of the British *Tachinidae* from the Fen District is at present so unsatisfactory that no record of any important species can be included here, but it is highly probable that the Fens with their many rare Lepidoptera would be certain to provide a few rare species of *Tachinidae*.

In the *Anthomyidae* the following can be recorded as occurring in this district of Cambridgeshire: *Hyetodesia obscurata*, Mg., and *H. umbratica*, Mg., *Alloeostylus flaveola*, Fln., *Hylemyia virginea*, Mg., *H. variata*, Fln., *H. lasciva*, Ztt., *H. pullula*, Ztt., *H. strigosa*, F., *H. puella*, Mg., and *H. coarctata*, Fln., *Lasiops ctenoctema*, Kow., *Anthomyia sulciventris*, Ztt., *Chortophila curvicauda*, Ztt., *Phorbia muscaria*, Mg., *P. cilicrura*, Rnd., and *P. trichodactyla*, Rnd., *Azelia zetterstedti*, Rnd., and *Lispe crassiuscula*, Lw.

Among the *Acalyptrate Muscidae* the following have occurred: *Cordylura umbrosa*, Mg., *Norellia liturata*, Mg., *Trichopalpus fraternus*, Mg., *Heteromyza atricornis*, Mg.,

Blepharoptera inscripta, Mg., *Trigonometopus frontalis*, Mg.,
Sciomyza obtusa, Fln., *Tetanocera laevifrons*, Lw., and *T. sylvatica*, Mg., as well as the commoner species. *Limnia marginata*, F., *L. unguicornis*, Scop., and *L. obliterata*, F., *Elgiva
dorsalis*, F., and *E. rufa*, Pnz., *Tetanops myopinus*, Fln., *Loxo-
desma lacustris*, Mg., *Ceroxys crassipennis*, F., *Chrysomyza
demandata*, F., *Trypeta jaceae*, Desv., *Tephritis corniculata*,
Fln., *Lauxania elisae*, Mg., *Piophila atrata*, F., *P. affinis*,
Mg., and *nigriceps*, Mg., *Anthomyza gracilis*, Fln., and *Diastata unipunctata*, Ztt. *Trimerina madizans*, Fln., taken at
Chippenham in heaps of cut sedge in the early spring, *Dis-
cocerina pulicaria*, Hal., *Philhygria picta*, Fln., and *flavipes*,
Fln., *Axysta cesta*, Hal., *Ilythea spilota*, Hal. *Lipara lucens*, Mg., common all over the Fens, the larvae living in
the reeds and causing the large conspicuous terminal gall.
This species may be easily bred, but is not often caught
because it drops from the reeds and feigns death upon the
approach of the collector. Other species of *Chloropidae* occurring in the Fen District include *Platycephala planifrons*, F.,
Haplegis divergens, Lw., which may be bred from the same
galls as *Lipara lucens*, the larvae living in the closely over-
lapping sheaths of the leaves, and *H. rufifrons*, Lw., which
at present has only been recorded from the Fens. *Oscinis
rapta*, Hal., and *Siphonella laevigata*, Fln., and *S. palposa*,
Fln., *Ochthiphila spectabilis*, Lw., *Borborus pallifrons*, Fln.,
B. pedestris, Mg., *B. longipennis*, Hal., *B. vitripennis*, Mg.,
and *B. geniculatus*, Mcq., *Sphaerocera eximia*, Coll., *Limosina
roralis*, Rnd., *L. curtiventris*, Stenh., *L. ochripes*, Mg., *L.
erratica*, Hal., *L. nivalis*, Hal., *L. rufilabris*, Stenh., and
L. vitripennis, Ztt., *Phora lugubris*, Mg., and *P. crassicornis*,
Mg. Species of *Borboridae* are exceedingly numerous in heaps
of cut sedge in the winter and early spring, and are very easy
to catch as they rarely attempt to fly, but if shaken out on
to a newspaper will only run towards the edge and can be
easily bottled; a splendid day's collecting of this sort has been
known to be had on a Christmas Bank Holiday, a fine bright

day after a slight frost. It may be of interest to record, that
of the 55 species of *Borboridae* in the British List, no less
than 50 have been taken in, or within a mile of, the borders
of Cambridgeshire.

Diptera have been but little collected in the Southern
District, though there are one or two noteworthy species that
have been captured. Mr F. Jenkinson in his garden at Cambridge has taken *Agathomyia collini*, Verr., which species
was described from specimens taken at Kirtling in this district,
and which has also been found in Herefordshire, the only
locality known at present outside Cambridgeshire; he has also
taken *Platypeza furcata*, Fln., and *P. dorsalis*, Mg., the rare
Mallota cimbiciformis, Fln., of which there is a second specimen in the Edinburgh museum, also taken at Cambridge; and
Systenus adpropinquans, Lw., a genus and species of *Dolichopodidae*, hitherto unknown in England and of which very
little is known on the Continent. Other rare species caught by
Mr Jenkinson are *Acletoxenus formosus*, Lw., *Stegana coleoptrata*, Scop., the only other recorded locality for this species
being the New Forest, *Anomoea antica*, W., a rare and very
handsome Trypetid, *Piophila latipes*, Mg., only otherwise
recorded from Herefordshire, *Chloropisca rufa*, Mg., *Phora
erythronota*, Strobl., and *P. abbreviata*, v. Ros.

Among other species taken in this district the following
are worthy of mention: *Docosia sciarina*, Mg., *Sceptonia
nigra*, Mg., *Polylepta splendida*, Winn., *Lasiosoma luteum*,
Mcq., *Asindulum flavum*, Winn., *Mycetobia pallipes*, Mg.,
Plesiastina annulata, Mg., *Anopheles maculipennis*, Mg.,
and *A. bifurcatus*, L., and very rarely *A. nigripes*, Staeg.,
Culex ornatus, Mg., *Goniomyia lateralis*, Mcq., *Molophilus
propinquus*, Egg., *Rypholophus nodulosus*, Mcq., *Tipula vari-
pennis*, Mg., *T. lateralis*, Mg., and *T. lutescens*, F., *Orthochile
nigrocoerulea*, Latr., *Pachygaster atra*, Pz., and *P. leachii*,
Curt., *Agathomyia antennata*, Ztt., *Platypeza infumata*, Hal.,
and *P. picta*, Mg., *Pipunculus fascipes*, Ztt., *Pipiza lugubris*,
F., *Chilosia vulpina*, Mg., *Melanostoma ambiguum*, Fln., *Syr-*

phus grossulariae, Mg., *S. nitens*, Ztt., *S. auricollis*, var. *nigri-tibius*, Rnd., and *S. punctulatus*, Verr., *Merodon equestris*, F., *Criorrhina floccosa*, Mg., *Idia lunata*, F., taken at Cambridge; *Hyetodesia lasiophthalma*, Mcq., and *H. basalis*, Ztt., *Phorbia neglecta*, Mde., *Homalomyia coracina*, Lw., and *H. polychaeta*, Stein, *Helomyza montana*, Lw., and *H. zetterstedti*, Lw., *Allophyla atricornis*, Lw., *Oecothea fenestralis*, Fln., *Dorycera graminum*, F., *Spilographa zoe*, Mg., *Trypeta colon*, Mg., *Urophora stylata*, F., *Tephritis formosa*, Lw., and *T. bardanae*, Schrk., *Urellia stellata*, Fuessl., *Toxoneura muliebris*, Harr., *Henicita annulipes*, Mg., *Anthomyza flavipes*, Ztt., *Diastata punctum*, Mg., *Philygria stictica*, Mg., and *P. interstincta*, Fln., *Drosophila transversa*, Fln., *D. phalerata*, Mg., *D. fenes-trarum*, Fln., and *D. obscura*, Fln., *Noterophila glabra*, Fln., *Camarota flavitarsis*, Mg., *Chlorops brevimana*, Lw., *Chloropisca glabra*, Mg., *Borborus nitidus*, Mg., *B. suillorum*, Hal., *B. roseri*, Rnd., *B. sordidus*, Ztt., *B. nigrifemoratus*, Mcq., *Sphaerocera monilis*, Hal., *Limosina lugubris*, Hal., *L. acutan-gula*, Ztt., *L. vagans*, Hal., *L. ferruginata*, Stenh., *L. halidayi*, Coll., *L. lutosa*, Stenh., *L. limosa*, Fln., *L. pumilio*, Mg., *L. sylvatica*, Mg., *L. scutellaris*, Hal., *L. clunipes*, Mg., *L. heteroneura*, Hal., *L. crassimana*, Hal., *L. quisquilia*, Hal., *L. fungicola*, Hal., *L. coxata*, Stenh., *L. spinipennis*, Hal., *L. minutissima*, Ztt., *L. mirabilis*, Coll., *L. nigerrima*, Hal., and *L. melania*, Hal.

It must be understood that this is not considered nor intended to be a complete list of the Diptera occurring in Cambridgeshire, as the commoner species are purposely omitted, and only those of interest or about which little is known are mentioned: among them will be found many noteworthy captures, and several species which are included, though in the British List, have not previously had the locality recorded in which they have occurred.

THE HYMENOPTERA OF
CAMBRIDGESHIRE.

By Claude Morley, F.E.S., &c.

So far as I am aware no attempt has ever been made to
work up the Hymenoptera of Cambridgeshire since the Rev.
Leonard Jenyns collected here during 1824—1849; his collection, named by Fred. Smith, is in the University Museum
at Cambridge, and it is from it that we must gather a general
idea of the bees, saw-flies and ichneumons of the county. As
long ago as 1797, the Rev. William Kirby traversed the fens,
and has left some account of the species he observed in his
Journal; and both Stephens and Curtis mention isolated
captures at Cambridge and Whittlesea Mere. To this ancient
and more or less unreliable material, I have only been able to
add a few records of the Parasitica by Bridgman in the *Trans.
of the Entomological Society*, 1882—89, and an occasional
mention by Saunders in his *Aculeata of the British Islands*,
1896. Several species have been sent to me thence by Cross,
Thornhill, Tuck, and Donisthorpe; and in 1902 I spent a
week at Wicken, but the weather was so inclement that but
few captures of even the commonest insects were effected.

Altogether only 173 species appear to have been noted in
Cambridgeshire, and of these 35 appertain to the Saw-flies,
68 to the Aculeates, and 70 to the Parasites; but many of the
older records must be received with due caution, especially
among the Tenthredinidae and Ichneumonidae. Of the former
Tenthredo solitaria, Scop., and *T. maculata*, Fourc. were taken
by Jenyns near Cambridge and at Wood Ditton respectively,

with *Allantus* (? *Macrophyia*) *rufipes*, Linn. at the latter
locality; I have found *M.* 12-*punctata*, Linn. at Wicken and
M. ribis at Burwell; *Poecilostoma submutica*, Cam. in Wicken
Fen and *Nematus fletcheri*, Bridg. at Burwell. *Cimbex femo-
ratus*, Cam., var. *Griffini*, is said to have occurred at Swaffham
Bulbeck and Ely in August; *Abia sericea*, Cam. on willows in
the Fens; and *Pamphilus sylvaticus*, Linn. at Wood Ditton.
Sirex noctilio, Fab. (*juvencus*, Fab.) is probably the species
bought by Jenyns from a man who found it commonly among
spruce firs at Fulbourn in June and July, 1837, though it is
called *S. duplex*, Shuck. (= *S. caeruleus*, Fab.), an American
species, never found in Britain—though it may possibly have
been imported. Its relative, the xylophagous *Xiphydria pro-
longata*, Geof. (*dromedarius*, Fab.) has been found upon
willows between Ely and Littleport.

Several interesting fossors, wasps and bees have been
noticed, of which not the least rare is *Myrmecina Latreillei*,
Curt., which Jenyns found at Swaffham Bulbeck in Sept. 1843,
and *Mutilla rufipes*, Latr. in sand-pits at Wilbraham Temple.
Saunders says the introduced ant, *Plagiolepis flavidula*, Rog.,
has occurred in Cambridge; and Thornhill has sent me *Crabro
vagabundus*, Panz. from Boxworth. *Gorytes campestris*, Linn.
was captured at the end of June in Bottisham Fen, and
Cerceris quinquefasciata, Rossi, near Stetchworth, with *Crabro
vagus*, Linn. at Ely, and *C. interruptus*, Deg. at Cambridge.
The rare wasp, *Odynerus Antilope*, Panz., and *Andrena apicata*,
Sm., have occurred to Perkins in Chippenham Fen, and the
same observer described a new bee, *Prosopis palustris*, from
Wicken [1], where I have captured *Sphecodes rubicundus*, Hag.;
and the very local bee, *Macropis labiata*, Fab., is found there
also at flowers of *Lysimachia vulgaris* and nesting in the
ground. I have confirmed Jenyns' record of *Nomada ochro-
stoma*, Kirb. by taking it at the Devil's Ditch; he says *Osmia
fulviventris*, Panz. occurs at Ely, with *Chelostoma florisomne*,
Linn. At Chippenham, Perkins has found *Stelis phaeoptera*,

[1] *Entomologists' Monthly Magazine*, 1900, p. 49.

Kirb., *Osmia pilicornis*, Sm., *O. bicolor*, Schk. *Chrysis ignita*, Linn. is common, and *C. viridula*, Linn. has been seen nesting in the wall of Swaffham Prior church.

Of the parasites, 58 belong to the Ichneumonidae and include such things as *Dinotomus lapidator*, Fab., which I recently brought forward as British; *Ichneumon dispar*, Poda, from Whittlesea by Curtis; the rare *I. lactatorius*, Desv., of which I found a specimen in the British Museum from the same locality; the fine *I. primatorius*, Forst., from Chatteris and Ely; and *Amblyteles microcephalus*, Steph., which appears to be quite unknown on the Continent and very rare in Britain. Jenyns bred *Cryptus ornatus*, Grav. from *Odonestis potatoria* in 1825, and Donisthorpe has sent me *Aptesis nigrocincta*, Grav. from an ant's nest at Wicken. Several species were first found here in Britain, such as *Ophion distans*, Thoms., *Sagaritis punctata*, Bridg., three *Mesochori*, &c. Thornhill has sent *Meniscus setosus*, Fourc. from Boxworth, and I have found *Xylonomus pilicornis*, Grav. at Wicken. Only seven species, including a *Bracon epitriptus*, Marsh., which I found at Wicken, of *Braconidae* are recorded, and the *Oxyura* and Chalcids fare worse with but two apiece, the latter being *Epicopterus choreiformis*, taken by Westwood at Gogmagog Hills, and *Cerapterocerus mirabilis*, Westw., from Cambridge; the *Evaniidae* is represented by *Foenus assectator*, Linn., which Thornhill has found at Boxworth, and which, with a great number of unnoticed species, is probably quite common.

THE MYRIAPODA OF CAMBRIDGESHIRE.

By F. G. SINCLAIR, M.A., of Trinity College.

THE number of species of Myriapods found in Great
Britain is not a very large one. This may in part be due to
the alterations that have been made in recent times in the
classification of the group. The two authors who have done
most work in the enumeration of the British species are
Newport and Leach. At the time when they wrote, the total
number of species known was not a quarter of those known
at the present day; and the characters which these writers
used in the determination of their species were not the same
as those which are utilized in more modern times. This
renders it a very difficult matter to come to a conclusion as to
how far the species known in this country extend on the
Continent, and whether we have in this country any species
altogether peculiar to it.

Though the range of species both of Chilopods and Diplopods is a wide one in one way, yet in another way they are
extremely local. This seeming contradiction is due to the fact
that one kind of habitat suits certain species and they are
rarely to be found in a place where the circumstances are
different. This is so much the case that Dr Karl Verhoeff
arranged the Diplopoda according to their habitat; thus:
(1) Diplopods living on heavy land; (2) on sandy land;
(3) under stones; (4) on leaves; (5) under bark; (6) on
plants; (7) cave Diplopods; (8) Alpine Diplopods; (9) foreign
Diplopods.

The Alpine and cave Diplopods are, of course, unrepresented in this country. The foreign Diplopods include those that come into this country on botanical specimens, vegetables &c. An example of this is the discovery of a *Scutigera* in Scotland by Gibson Carmichael. It is not impossible that this animal may really have been bred in this country, but it was most probably imported. A specimen of *Polyxenus* has been described as having been found in Phœnix Park in Ireland, by Carpenter. The geographical distribution of this species renders it probable that it was brought into the country on some plant or shrub.

Assuming that these animals were so imported, they would be examples of what Verhoeff means by foreign (fremdlinge) in his classification by habitat.

Although Myriapods are very local in their habits and may usually be found in one spot, so that one would be almost sure to find *Cryptops hortensis* under stones or flower-pots in almost any garden: or *Julus terrestris* under the bark of trees; yet there is no doubt that Myriapods do sometimes make migrations in large numbers for considerable distances.

The author before mentioned, Verhoeff, describes two instances of this which are very interesting. In 1878 there was a migration of *Brachyjulus unilineatus* in Russia, and the animals were in such numbers that they actually stopped a train. Their crushed bodies made the line so greasy that the engine was unable to proceed.

Again, in West Germany, in June 1890, a train was stopped in the same manner by a migration of *Schyzophylum sabulosum*. There were two varieties of this species in this migration, one known as *S. bifurcatum*, the other as *S. punctulatum*. The latter, however, was in much smaller numbers. An example of a more local migration is supplied by Verhoeff in the case of *Brachydesmus superus*, which live in dry seasons beneath the earth and in rainy weather are found above the ground: in the latter case they make short migrations.

The seasons at which Myriapods breed are the spring and

summer. There seems, however, to be rather a wide margin
of difference in different localities for the breeding season.
I found that in Cambridge and the surrounding localities the
Julidae breed in June and July, *Polydesmus* at the same time,
Lithobiidae somewhat later (in July, August, and September).
A friend of mine in South Wales informs me that he has found
Polydesmus, Glomeris, and *Julus* breeding as early as March,
while in the extreme north of Scotland I have found the eggs
of *Polydesmus* as late as October. All the Diplopods that
I have observed myself form nests in which they lay their
eggs, while Lithobius, which is the only Chilopod that I have
had an opportunity of observing, lays its eggs singly.

A list of the British species, noting the other countries in
which they have been found, will enable us to distinguish
those which inhabit the more northern countries from those
which are common in the warmer climates and only occur
rarely in the northern latitudes.

Of the genus *Scolopendrella* only one species has been
found in Britain: *Scolopendrella immaculata.* This species
has also been recorded in Russia and Germany.

Of the *Pauropidae* there are two British species: *Pauropus
huxleyi,* England, Germany, and Russia, *Pauropus peduncu-
latus,* England, Germany, and Russia.

Of the *Pselapsognatha* one species, *Polyxenus lagurus,* has
been found in Ireland by Carpenter. This species belongs to
the south of Europe and becomes rarer as the more northern
climates are reached, which renders it probable, in my opinion,
that the solitary specimen found was imported from another
country.

The Chilognatha have been divided by Pocock into *Oniscomorpha, Limacomorpha,* and *Helminthomorpha.*

1. Of the *Oniscomorpha* the only representative is
Glomeris marginata, Leach, which is fairly common in the
South of England and Wales, and has been found in Ireland
by Carpenter. It has also been found in France, Germany,
and Italy. I have not found any record of its occurrence in

the more northern parts of England or Scotland, though it occurs in Denmark.

2. The *Limacomorpha* are unrepresented in this country.

3. The *Helminthomorpha* comprise the greater number of the British *Chilognatha*.

Polydesmus complanatus. A widely-distributed species, common in England, Ireland, Wales, and Scotland. Found all over the Continent, in Madeira and the Azores.

Brachydesmus superus. Hungary, Austria, and Germany. I have found it in the North of Scotland, in Essex, Devon, and Wales.

Craspedosoma polydesmoides, Leach, Syn. *Atractosoma bohemicum*, Latzel. Austria, Germany, and England.

Craspedosoma rawlinsii, Leach. Hungary (rare), Austria, Mid-Europe; seems not to be known in France.

Atractosoma latzelii, Verhoeff. South England. I have received specimens from the neighbourhood of Plymouth, and found specimens in Essex. This species does not seem to have been recorded from any country but England.

Of the genus Julus, Leach gives the following species :—
J. londinensis, J. niger, J. terrestris, J. sabulosus, J. punctatus, J. pulchellus, and *J. pusillus.* As has been before mentioned, there is a great deal of difficulty in identifying some of these with the continental species. *J. sabulosus* is, according to Latzel, synonymous with *J. terrestris* of Linnaeus, and *J. rubripes* of C. Koch. I have had specimens from Wales, Devon, Cumberland, and Perthshire. The specimen I received from the latter locality was the variety named by Verhoeff *J. bipunctatum.*

J. londinensis, Leach, is identified by Verhoeff with a species occurring in Germany. He has obtained specimens from the neighbourhood of London as well as from various parts of Germany and believes them to be the species which Leach described as *J. londinensis.* Verhoeff's species, however, does not correspond with Leach's description in two points. Leach describes the praeanal segment as sub-mucronatum, while

Verhoeff's species has no trace of an anal spike. Then as to their size, Leach's specimen was 2½ inches long, while the largest specimen examined by Verhoeff is only 38 mm. Fanzago has described *J. londinensis*, Leach, as found in Italy, but I am informed by a friend that his species is not Leach's but probably synonymous with *Diplolulus apenninorum*, Brühl. *J. niger* is, according to Latzel, identical with his *J. scandinavius*, the *micropodiulus ligulifer* of Verhoeff, and *J. terrestris* of Meinert and others. It has been found in Germany, Austria, and Scandinavia. I have received specimens of it from Devonshire and Cornwall. *J. terrestris*, Lin., is, according to Latzel, synonymous with *J. fallax* of Meinert and others. It is a widely-distributed species on the Continent and common in this country.

J. punctatus. Latzel identifies Koch's *J. punctatus* with his (Latzel's) *J. pelidnus* but does not believe it to be the same as Leach's *J. punctatus*.

J. pulchellus is identified by Latzel with *Blanjulus venustus*, Meinert. It is a common species both on the Continent and in this country. *J. pusillus* he identifies with *J. boleti* of Stein and Porath, and *J. stuxbergi* of Fanzago and Berlese. It has been found in Germany and Italy.

Blanjulus guttulatus is a widely distributed species; it is fairly common in Cambridgeshire.

In addition to these species Newport gives *J. sandivicensis*, and it is very doubtful what species it corresponds to. In addition to the species of Julidae already mentioned, I have found two species common in England and Ireland; one of these is *J. albipes* which Verhoeff gives as common in England. Latzel queries if it is not a synonym of *J. fallax*, but there is no doubt that Verhoeff is right in holding it to be a distinct species. This species which I have received from Devon, Cambridge, and South Wales, I believe to be identical with Newport's *J. sandivicensis*. The other species, which I have found abundant in county Wicklow, in Ireland, is the species described by Latzel as *J. longabo*. Verhoeff gives it as a

synonym of *J. fallax* but I think the two species are distinct.
There is a marked difference in the size of the auxiliary
copulatory organs in relation to the size of the animal, and the
form of the metamorphosed first pair of legs in the male differs
considerably in the two animals. In the specimens which
I have examined— about 50 of *J. longabo* and 200 of *J. albipes*
—there is a slight difference in the form of the anal spike.

Carpenter records *Julus luscus*, Meinert, in Ireland. Verhoeff describes a species of Julus which he calls *J. britannicus*
as occurring in England and resembling *J. luscus* very closely.
It seems to me not improbable that Carpenter's species was
really Verhoeff's *J. britannicus* as the two species resemble
one another so closely, and *J. luscus* has not, so far as I know,
been found in Great Britain.

The Chilopoda have been divided by Pocock into *Geophilomorpha, Scolopendromorpha, Craterostigmomorpha, Lithobiomorpha* and *Scutigeromorpha*. Of these, all except the *Craterostigmomorpha* are represented in Britain.

1. Of the *Geophilomorpha* there are the following
species: *Geophilus longicornis*, Leach; synonyms, *G. elec-
tricus*, Gervais; *Arthromalus longicornis*, Newport; *A. flavus*,
Newport; *A. similis*, Newport. This species is found in
Austria and Germany, and varieties, or nearly allied species,
in Portugal and the Azores. I have found it fairly common
in Cambridgeshire. *Arthromalus . carpophagus*, Newport;
synonyms, *Geophilus condylogaster*, Latzel; *G. sodalis*, Meinert;
found in England, Ireland, and Austria. *Geophilus humili*,
Newport; I believe this to be synonymous with *Scolioplanes
acuminatus*. *G. subterraneus*, Newport; I am unable to say
whether this is a distinct species. *Scolioplanes acuminatus*,
fairly common in Cambridgeshire, and England. It is also
found in Germany and Austria. *Scolioplanes maritimus*,
Leach. Verhoeff considers this as a race or a variety of *Sc.
acuminatus* or a variety of *Sc. crassipes*. *G. breviceps*, Newport,
synonymous with the above (*Sc. crassipes*).

2. The *Scolopendromorpha* are represented in this country by a single form *Cryptops hortensis*, synonymous with *C. savigny*. It is common in gardens.

3. Of the *Lithobiomorpha* the British species are *Lithobius variegatus*, Newport. Distinguished by the curious annular markings on its limbs. I have had this species from Devon, Cornwall, and Ireland. *L. forficatus*, Leach; a common species both in this country and on the Continent. It is synonymous with *L. laevilabrum*, Leach, *L. vulgaris*, Leach, *L. Leachii*, Newport. A very common species found over most of Europe and even in North and South America.

L. pilicornis; according to Pocock, this species is synonymous with *L. sloanii*, Newport, *L. longipes*, Porath, and *L. galathea*, Meinert. It has a wide distribution in southern countries, having been found in Madeira, Morocco, and the Azores.

I have had specimens from Cornwall but have not been able to find it in the North of England or Scotland.

L. melanops, a very small species. According to L. Koch it does not occur in Germany. *L. microps*, another small species which, according to Carpenter, is found in Ireland. It also occurs in Denmark. *L. crassipes*, I have got specimens from Dartmoor and Cornwall.

4. The *Scutigeromorpha* are represented by a single species *Scutigera coleoptrata*. This species was found in Scotland by Gibson Carmichael, but the occurrence of a species which is confined to hot climates, so far north, leads one to believe that it may have been imported. I have however had one sent me for examination from Jersey, and as this was a young specimen it was probably bred there.

Of the Myriapods of Cambridgeshire in particular, the information is very scanty. I have myself collected specimens in the part of the county surrounding Cambridge and extending to Royston and the boundary of Essex, towards Heydon and Audley End, but I cannot say that even in this part of the county my examination has been at all exhaustive. I have no doubt that many more species could be discovered than I have been able to find. I have myself collected the following species.

Julus albipes, on bark of willow-trees about Grantchester, and in the fields about Whittlesford Common.

J. fallax, fields about Cambridge and at Newton, fairly common. *Blanjulus guttulatus*, common in dried leaves, under stones or logs of wood in fields about Hauxton, Foxton, Newton and other villages. *Polydesmus complanatus*, under logs of wood, not very common near Cambridge. I find that on clay lands they are more common than on lighter soils. *Lithobius forficatus*, common under logs of wood and stones near Cambridge and the neighbourhood. This is perhaps the commonest of the English species.

Lithobius melanops, a very small species common in gardens. *Cryptops hortensis*, not very common in Cambridgeshire. It is found in gardens, but in my experience it is more abundantly found in heavier soils than that of Cambridgeshire. *Scolioplanes acuminatus:* this species I have found fairly common in gardens and fields about Cambridge. It is found under logs and dried leaves and more especially just beneath the surface of the soil. *Geophilus longicornis*, under logs and stones, and like the last-mentioned species, just beneath the surface of the soil. The heath at Royston is a good place both for it and the last-named species.

Now there are several other species for which it would be worth while to search, considering, as I have already said, that the county has never been systematically searched for Myriapoda. *Pauropus huxleyi* might be found under dried leaves. *Glomeris marginata* is, I think, from what I know of its distribution, unlikely to be found, but *Atractosoma latzelii* would be a probable find and might be searched for in moss and dried leaves, especially in any wooded part such as the woods about Madingley. Verhoeff's *Julus britannicus* might be looked for in moss near the banks of streams. *Geophilidae* might be searched for on Royston Heath and in the uncultivated part of the Fen. I think it probable that new species might be found. *Lithobius variegatus* might be found under stones and logs on light sandy soil, and I think it probable that the uncultivated Fen might yield this species. I have myself found it on peaty soil. It is not improbable that *J. sabulosus* might be found, although I think that it is more common in more southern parts of England; still I have had it sent to me from much more northern parts of England than Cambridge. According to Verhoeff's classification it is found on plants, flowers of Ranunculus, sandy soil and leaves. *Lithobius crassipes* I have only found in the South of England, but if it is to be found in Cambridgeshire, it would be under stones in the Fen.

THE ARACHNIDA OF CAMBRIDGESHIRE.

By C. WARBURTON, M.A., Christ's College.

THE four British orders of land Arachnida, (1) Chernitidea or false-scorpions, (2) Araneae or spiders, (3) Phalangidea or harvestmen, and (4) Acari or mites, are all well represented in the neighbourhood of Cambridge. But while the Cambridge list of spiders is a respectable one, comprising more than 180 out of about 500 British species, and a certain number of the comparatively few British species of harvestmen and false-scorpions have been locally recorded, hardly any attempt has been made to grapple with the large and important group of the mites, which probably, in the number of species, and most certainly in the number of individuals, would far out-number all the other local Arachnida put together.

A few words will suffice with regard to the local false-scorpions or harvestmen, but there are points of interest in connection with the spiders which call for a somewhat larger notice.

(1) The **Chernitidea** or false-scorpions are creatures so small in size, and so retiring in habit, that it is quite possible for an observant person to pass years without ever seeing a specimen, unless he is specially searching for these animals or for other small creatures which affect similar haunts. It is just possible he may have noticed examples clinging to the legs of house-flies, or come across a "book-scorpion" between the leaves of some disused volume. Such chance occurrences are rare, however, and anyone who wants to make acquaintance with these creatures will do better to search diligently under

stones or bark, or to gather moss and shake it out over a sheet of paper.

When found, they are certainly very quaint objects, especially when examined with a lens. No English species measures more than ⅛ in. in length, and some are much smaller. They are usually straw-coloured or brown, and the appropriateness of the name false-scorpion is at once apparent; their long chelate pedipalps are remarkably scorpion-like, but the "tail" or postabdomen with its terminal sting is wanting. Their usual gait is slow and deliberate, but they can scramble away with considerable speed—most readily with a lateral or retrograde motion. They are predaceous, feeding on small insects, and I have detected them devouring Psocidae.

Six genera and twenty species are recorded as English, and of these four genera and seven species have occurred at or near Cambridge, and so small a list may be given in full:—

GROUP I.   Four eyes.

Chthonius orthodactylus, Leach.
Chthonius rayi, L. Koch.
Chthonius tetrachelatus, Preyssler.
Obisium muscorum, Leach.

GROUP II.   Two eyes.

Roncus lubricus, L. Koch.

GROUP III.   No eyes.

Chernes nodosus, Schrank.
Chernes phaleratus, Simon.

(2) The **Phalangidea** or harvestmen are usually classed as spiders by the uninitiated. Some of them are very familiar objects, with excessively long and brittle legs, and comparatively small bodies, though this disproportion by no means always holds good. In reality they are easily distinguished from spiders by the fact that the whole body is in one piece

—without any waist—and that it shows distinct traces of segmentation. There are, of course, many other points of difference, and it may be mentioned that a phalangid always has two eyes only, whereas the normal allowance for a spider is eight. Moreover phalangids do not spin webs. They are found on walls, under stones, and among grass and herbage, and are predaceous animals. Twenty-four species are British, belonging to nine genera. Of these, four genera and eight species have been recorded in Cambridgeshire. They are given below:—

Phalangium opilio, Linnaeus.

Phalangium parietinum, De Geer.

Phalangium saxatile, C. L. Koch.

These are fairly large species, with moderately long legs, found on walls and under stones.

Liobunum rotundum, Latreille, the species with small dark body and immoderately long legs, found on walls or running among herbage.

Nemastoma lugubre, Müller. A short-legged species, black, with two white spots, common in swampy places among grass.

Oligolophus morio, Fabricius.

Oligolophus ephippiatus, C. L. Koch.

Oligolophus spinosus, Bosc.

These species have a large body, and moderately long and rather strong legs. They are found under stones and among herbage.

(3) Of all the **Arachnida** the spiders are far the most interesting in their habits, popular prejudice notwithstanding, and anyone who undertakes the study of these creatures is sure to find it fascinating. They present a very great variety in their mode of life. The spinners of the circular snare are familiar objects enough, and so are the less attractive sheet-web weavers of cellars and out-houses, but the wolf-spiders which hunt their prey through the herbage—the mothers with the newly-hatched young on their backs—and the jumping

spiders with their remarkable acrobatic feats on perpendicular surfaces, and the multitudinous weavers of irregular webs, are not so generally known. Their spinning operations, whether in spreading snares, constructing retreats, or weaving egg-cocoons, no doubt furnish the greatest field of interest, but apart from these there are many points in their economy which are highly curious, as, for instance, the aeronautic expeditions of the migrating young, and the extraordinary dangers encountered by the males of some species during the breeding season.

Among the Cambridge spiders there are many notable forms, and some that have not yet been found elsewhere.

Our nearest English congener of the trap-door spiders of southern Europe, *Atypus piceus*, is to be found on the Devil's Dyke. It certainly constructs no trap-door, but it forms a deep burrow, lined with silk, which is continued into a purse-like structure in the herbage above the level of the ground. *Tegenaria domestica*, the largest English spider—at all events as regards the span of its legs—is common in cellars and out-houses. The water-spider, *Argyroneta aquatica*, abounds in the local streams and ditches. Nearly all the English species of *Theridion* occur in Cambridgeshire, and some of the rarer Epeiridae are not uncommon. But the chief characteristic of the local aranean fauna is derived from the adjoining fen-land, and in Wicken Fen particularly are to be found several species very rarely met with elsewhere.

The handsome wolf-spider, *Lycosa farrennii*, was long thought to be peculiar to Wicken, and still its only other locality is in the outskirts of Paris. One of our finest jumping-spiders, *Marpissa pomatia*, stood for many years in the British list on the strength of a single female taken in the North of England, but both sexes occur with tolerable frequency among the sedge of Wicken Fen. And besides these special rarities there are a variety of fen forms which would prove a welcome addition to the collection of any arachnologist who had hitherto worked chiefly on higher ground.

A brief analysis of the local spider fauna will probably be more useful here than a complete list of species.

The classification of the Araneae is by no means at present in a crystallised condition, but according to our view thirteen families are represented in the neighbourhood. Five of these, however, account for only eight species. *Atypus piceus*, mentioned above, is the only example of the Atypidae; the little brick-red six-eyed spider, *Oonops pulcher*, is the sole representative of the Oonopidae; of the Dysderidae—six-eyed spiders of fair size, living under stones or bark—the three commoner species are plentiful around Cambridge; the pretty little *Ero furcata* is the only member of the Mimetidae, while of the Pisauridae, *Pisaura mirabilis* is common, and the large and handsome *Dolomedes fimbriatus* has been recorded.

The remaining eight families call for a somewhat more extended notice.

DRASSIDAE. These are elongate flat spiders, living under stones or bark or in curled leaves, or in some cases running freely on the ground. Their colouration is usually dull, the majority being brown or mouse-coloured, and a few jet-black. There are more than twenty local species. The large mouse-coloured spider, so often seen lurking under a stone, is *Drassus lapidosus*. *Clubiona* has several species, and the handsome *C. corticata*, more elaborately marked than most of the group, is common under bark. *C. subtilis* is very common in the fens, while the rare *C. lutescens* is taken at Wicken. The jet-black Drassids belong to the genus *Melanophora*, and Wicken is one of the few localities for *M. lutetiana*.

DICTYNIDAE. A small group of "cribellate" spiders, the female possessing an extra spinning organ, the "cribellum," correlated with a comb on the hind leg called "calamistrum." *Amaurobius* lives in cellars and out-houses, while *Dictyna* is found among herbage.

AGELENIDAE. This not very extensive group of sheet-web spiders includes some very familiar examples. The large *Tegenaria domestica* and the smaller *T. derhamii* are common in houses, while everyone knows the web of *Agelena labyrinthica*, to be met with so frequently on the banks of ditches. The well-known water-spider, *Argyroneta aquatica*, common in the neighbourhood, belongs to this family, as does also the handsome *Textrix denticulata*, a welcome addition to most collections, though not rare at Cambridge under the loose bark of trees.

THERIDIIDAE. The great majority of British species belong to this family, and three of its sub-families, Theridioninae, Erigoninae and Linyphiinae, are very well represented about Cambridge.

The Theridioninae, of which the principal genus is *Theridion*, include the prettily variegated, somewhat globular spiders, whose irregular webs are to be found on almost every holly bush. Perhaps the most beautiful example, *Theridion formosum*, is in some seasons extremely common, though ordinarily seldom met with.

The Erigoninae comprise the multitudinous tiny black or brown spiders, sometimes called "money-spinners." Some of these are very curious when examined under the microscope, the heads of the males in certain species being furnished with variously shaped prominences on which the eyes are mounted. An extreme example is *Walckenaera acuminata*, the height of whose slender eye-turret equals the length of the cephalothorax. One of the smallest known spiders, *Panamomops bicuspis*, about $\frac{1}{20}$ in. in length, is a local rarity. *Wideria warburtonii* has not yet been found elsewhere. On fine days in the autumn many of these spiders, together with the newly-hatched young of larger species, may be observed on iron railings busily occupied in aerial migration. Standing on tip-toe, so to speak, the spider raises its abdomen and emits a slight flake of silk, which is gradually drawn out by light currents of air till the pull is sufficient to support the weight of the spider, which then lets go its hold and floats gently away.

The Linyphiinae include those spiders which are conspicuous objects hanging beneath their hammock-like webs in bushes and herbage, and also a multitude of forms only to be discovered by searching carefully among the grass and sedge. Among the latter are many rare fen species, though the group is not one which can lay claim to any particular beauty.

EPEIRIDAE. These are the spinners of the familiar circular snare, and include the common garden spider *Epeira diademata*. More than twenty species are locally recorded, and they include the handsome *E. pyramidata*, which, however, is rare, and *E. sclopetaria*, which is tolerably plentiful. It is in this group that the male runs the singular risk of being summarily trussed up and devoured by the female if she chances to be hungry when he is paying his attentions, and he always approaches her snare with a hesitation which is perhaps scarcely to be wondered at. The spinning operations of the Epeiridae are much aided by the extreme mobility of the abdomen, which is united to the cephalothorax by a very narrow pedicle.

THOMISIDAE. From their latigrade motion, and the general crab-like appearance of many of these spiders, they are popularly known as crab-spiders. About fifteen species are found in Cambridgeshire. Among the rare Wicken species are *Xysticus pini*, *Oxyptila trux*, *O. atomaria* and *Thanatus striatus*, while *Oxyptila simplex* is found on Fleam Dyke.

LYCOSIDAE. These are the "wolf-spiders," examples of which may be often seen in vast numbers running over the ground among dead leaves and herbage. In the breeding season the females carry the egg-cocoon on all their expeditions, attached by silken threads to their spinnerets, and when the young hatch out they mount on the mother's back. Some make a silk-lined burrow in the ground. The most interesting local form is *Lycosa farrennii*, which is extremely common in Wicken Fen though not found elsewhere, unless Simon is correct in thinking that he has recently met with it near Paris.

ATTIDAE. Jumping spiders.

The most familiar example is *Salticus scenicus*, a zebra-marked species, common on walls and fences. It trails a silken line to anchor it in case of accidents, wherever it goes, and stalks insects warily till it comes within leaping distance, pouncing upon them with remarkable precision.

The rare *Attus pubescens* is not uncommon on the gray walls of the colleges. *Marpissa pomatia* has already been mentioned as one of the rarities of Wicken Fen, but the allied *M. muscosa* is very seldom taken in Cambridge.

Two species of *Heliophanus* and one of *Euophrys* frequently occur in herbage.

Besides their habit of progressing by leaps—very unequally exhibited by different members of the group—these spiders are remarkable for their eyes, four of which are situated on the forehead with their axes directed forward, two of them being generally very large and noticeable. They spin no web, and the great mobility between the cephalothorax and abdomen, instead of facilitating their spinning operations, as in the Epeiridae, is chiefly of use in enabling them to direct this powerful battery of eyes with great rapidity upon their prey.

(4) Of the local **Acari** or mites there is, as has been said, but little to record. The minute size of these creatures and the lack of any satisfactory monographs, except in the case of one or two groups, are serious obstacles in the way of the collector. The only acarid family which has attracted any attention from Cambridge naturalists is that of the ORIBATIDAE or beetle-mites, and the search for them has been so recently undertaken that an attempt to give anything like a complete local list is decidedly premature. Perhaps a brief statement of what has been done so far may serve to direct attention to a group of creatures very interesting in themselves, and possessing certain distinct advantages from a collector's point of view.

In the first place the group has been ably monographed by Michael in the publications of the Ray Society, so that the labour of identification is comparatively small. Then the creatures are entirely microscopic—that is to say they have to be examined from first to last under the microscope—and in this respect they are more convenient to deal with than creatures too large as a whole for microscopic objects, but necessitating microscopic examination of anatomic details. Finally, there are decided advantages in connection with the search for Oribatid mites. Most of them live under bark or in moss and lichen, but their small size makes it almost useless to try and collect individual specimens out of doors, and the best plan is to bring home bags full of moss and pieces of loose bark and to shake them over white paper, when the tiny creatures soon betray themselves by wandering about among the specks of inanimate matter shaken out with them, and the collector can indulge in the pleasures of the chase on his own study table with all his apparatus about him. To one who is fond of working with the microscope and is skilful in manipulating small objects, the Oribatidae are certain to prove a very attractive group.

The search during three or four months, and these at the least favourable time of the year, has revealed some forty-five Cambridge species of Oribatidae out of about 100 British species, and these are divided among fifteen genera. In size they range from something over a millimeter to ·3 mm. Some of them are extremely elegant in form, while anything more grotesque than others it is difficult to imagine.

Most of the species so far discovered belong to the genera *Oribata* and *Nothrus*. Mites of the genus *Oribata* are round or oval, with hard polished integuments, which have given to the group the popular name of beetle-mites. About a dozen species are already included in the local fauna. The genus *Nothrus* is very different in appearance. The integuments are leathery and rough, and often bristling with the oddest array of protuberances and spines. Six species have been recognised in the neighbourhood of Cambridge.

The following must be regarded as a preliminary list of the Cambridge Oribatidae. There is little doubt that it will be considerably extended even during the next few months. A few species of doubtful identification are omitted :

Pelops acromios, Hermann.

Pelops fuligineus, C. L. Koch (*P. laevigatus* in Brit. Oribatidae).

Pelops phaeonotus, C. L. Koch.

Oribata globula, Nicolet.

Oribata orbicularis, C. L. Koch.

Oribata piriformis, Nicolet.

Oribata setosa, C. L. Koch.

Oribata ovalis, C. L. Koch.

Oribata avenifera, Michael.

Oribata quadricornuta, Michael.

Oribata cuspidata, Michael.

Oribata lucasi, Nicolet.

Oribata fusigera, Michael.

Oribata parmeliae, Michael.

Oribata lapidaria, Lucas.

Scutovertex maculatus, Michael.

Scutovertex sculptus, Michael.

Cepheus tegeocranus, Hermann.

Cepheus latus, Nicolet.

Tegeocranus velatus, Michael.

Serrarius microcephalus, Nicolet.

Liacarus ovatus, C. L. Koch (*Leiosoma ovatum* in Brit. Oribatidae).

Notaspis similis, Michael.

Notaspis exilis, Nicolet.

Notaspis tibialis, Nicolet.

Notaspis oblonga, C. L. Koch.

Notaspis lucorum, C. L. Koch.

Notaspis splendens, C. L. Koch.

Damaeus verticillipes, Nicolet.

Damaeus clavipes, Hermann.

Damaeus geniculatus, Linnaeus.
Damaeus auritus, C. L. Koch.
Hermannia scabra, C. L. Koch (*H. nodosa* in Brit. Oribatidae).
Hermannia arrecta, Nicolet.
Hermannia bistriata, Nicolet.
Neliodes theleproctus, Hermann.
Cymbaerimaeus cymba, Nicolet.
Nothrus targionii, Berlese.
Nothrus sylvaticus, Nicolet.
Nothrus palustris, C. L. Koch.
Nothrus horridus, Hermann.
Nothrus biverrucatus, C. L. Koch.
Nothrus bicarinatus, C. L. Koch.
Hypochthonius rufulus, C. L. Koch.
Hoploderma magnum, Nicolet.
Hoploderma dasypus, Dugès.

THE CRUSTACEA OF CAMBRIDGESHIRE.

By W. A. CUNNINGTON, Christ's College.

In a survey of the Crustacea of Cambridge and its neighbourhood, we have at the outset our task greatly lightened
by the inland situation of the town.

The great majority of the members of the group are inhabitants of the sea, a certain proportion inhabit fresh water,
and a minority are adapted for a life upon land. It follows
then that, of the immense number and variety of forms which
together constitute the group, we have but comparatively few
to take into consideration. The common crab or lobster,
typical representatives of the class, are of course marine, and
we are left with forms the majority of which are quite small,
often almost microscopic.

While there is no reason to suppose that the terrestrial
Crustacea are strikingly different from those common in other
parts of our island, the fresh-water types from Cambridge,
surrounded as it is by the old fen country, might well be
numerous and interesting. Unfortunately we have here to
face a great lack of information. It is indeed remarkable
that a district containing a great University should have
been so little exploited, at least in this direction, but the
fact remains[1]. Almost the only work of reference is a
manuscript "Catalogue of the Insect Fauna of Cambridgeshire" by the Rev. Leonard Jenyns (afterwards Blomefield),

[1] The Cambridge Natural History Society has recently taken up this
faunistic question, and is endeavouring to obtain complete lists and collections of the local fauna.

which he presented to the Cambridge University Museum in
1868.

It has been thought that the addition of certain facts concerning the structure and habits of the Crustacea occurring
in the district, would render this account of more interest
to the general reader, and it has accordingly been done. The
extremely local and variable distribution of many of the
forms to be mentioned must also be noticed. A species,
which is not to be found at one time, may appear in enormous
quantities shortly after. The reverse is of course equally
true, so that it by no means follows that every species will be
found in the particular spot mentioned.

It will perhaps be most satisfactory to take a survey of
the various Orders which constitute the class Crustacea, and
so to consider in turn the principal characteristics of the.
more important forms which come under consideration.

In the first instance it will be sufficient for our purpose
to divide the Crustacea into the two great groups of Ento-
mostraca and Malacostraca. The former includes the smaller
and more simply organized forms and, in the present instance,
the great majority of those we have to consider. The group
Malacostraca contains generally speaking the larger and higher
forms, most of which are inhabitants of the sea.

ENTOMOSTRACA.

PHYLLOPODA. The Order Phyllopoda, which contains the
lowest of all Crustacea, contains also a large number of most
varied forms which are characteristic of inland waters, and
well represented in the Cambridge district.

It is not indeed the most primitive types,—the Branchio-
poda—which occur here, but rather members of the more
specialised sub-order Cladocera, the so-called "water fleas."
With very few exceptions, the whole of the Cladocera live in
fresh water, and so may well be abundant in a country of
pond, ditch, and fen. The common *Daphnia*, which is a
most familiar object to all who have studied pond life, may

serve as a typical Cladoceran. With the exception of the
head, the creature is laterally flattened and enclosed in a
bivalve shell, while locomotion is effected by the powerful
strokes of the large second antennae. *Daphnia pulex* is one
of the commonest and best known species, and occurs in this
district in great abundance. Very common also is *Simocephalus sima*, which was formerly included under the heading
Daphnia. Closely allied we have also *Daphnia magna* and
D. longispina.

All these forms are active swimmers, but the style of
swimming affords a ready means of distinction between the
genera *Daphnia* and *Simocephalus*. The members of the
former genus swim in a vertical position, head uppermost,
by constant beating of the large biramous second antennae.
Simocephalus, on the contrary, swims on its back with some
vigour, but never for long without coming to rest. Since it
has been demonstrated that the movements of the swimming
antennae are in both cases almost identical, we can only
believe that the centre of gravity must be quite differently
situated in the two genera.

A further very characteristic attitude of *Simocephalus* is
that which it assumes on coming to rest, when it hangs on
to water-weeds or the like, by means of minute hooks at the
end of the second antennae. This position is evidently a
favourable one for respiration and the taking in of food, for
both processes are effected by a regular beating of the five
pairs of legs, to produce a circulation of the water between
the valves of the shell. Some idea of the importance of this
movement of the legs may be gathered from the fact that
they usually perform no less than 300 beats within the minute.
It is easy to understand that all light particles of organic
matter would be carried by the current of water within reach
of the mouth appendages, and that the food of such forms
thus consists entirely of small unicellular organisms with both
animal and vegetable detritus.

The process of shedding the skin, which is periodically

gone through by all Crustacea, has been observed to take place in these forms with surprising frequency. In the last mentioned species, *Simocephalus sima*, the moult takes place with considerable regularity, every four or five days in the case of the adult, and probably still more often in young and growing specimens.

The relations of these Cladocera to the surface-film of water are worth attention. In cases where the creatures have been kept captive, it is no infrequent thing to find that certain specimens have become entangled in the surface-film and are unable to sink below the surface. Whether such an occurrence is ever to be met with in nature, and if not, why it is not the case, are points which remain to be investigated. In this connection, however, it is interesting to note that one of the Daphnidae—*Scapholeberis*—deliberately makes use of the surface-film for support, attaching itself by the modified hairs on the ventral margin of its shell.

Parthenogenesis, or reproduction without fertilisation, during certain periods of the year, is a characteristic feature of Cladocera, and these common forms are among the most prolific of them all. *Daphnia pulex* may have frequently as many as 40 eggs at a time in the brood chamber, and only four or five days are necessary for the complete development. As of course each offspring soon becomes capable of continuing the process, it will be seen that the rate of increase must be most prodigious. A calculation, based on numbers below the average, gives the startling result that the descendants of a single female, after 100 days, would number one and a half billion. After fertilisation, which takes place in the autumn, and occasionally at other periods of the year, the female produces a small number of eggs which are enclosed in a protective case, the so-called ephippium.

A few other forms of Cladocera are perhaps worthy of notice. *Chydorus sphaericus* is one of the commonest of all, occurring in most of the ponds and ditches, though from its small size it is easily overlooked at first sight. It is a tiny

round creature with comparatively small second antennae and
a compound eye, which is very inconspicuous when compared
with the large structures in the Daphnidae. *Eurycercus
lamellatus* is also common, and from the characteristic saw-
like nature of. the abdomen, is easily distinguishable. It
has been taken in the river Cam, as well as in various streams
and ditches. The fen ditches near Reach have proved a
happy hunting-ground for Cladocera. In addition to most
of the forms already mentioned, we find species of *Alona*,
Pleuroxus, and *Ceriodaphnia*, with an abundance of the
highly characteristic *Polyphemus pediculus*. This latter, as
its name denotes, is possessed of a relatively enormous compound eye, and when swimming in the water, this large black
spot, together with the brownish U-shaped intestine, are alone
conspicuous, owing to the transparency of the remaining
parts.

Finally, an interesting form is *Macrothrix hirsuticornis*,
taken in the village ponds at Barton. The animal in question
would seem to come nearest to the specific definition of
hirsuticornis, but is unaccountably large. Even if merely
M. hirsuticornis, the capture is interesting, as the species is
only recorded from four other localities in the British Isles.

OSTRACODA. A number of these small crustaceans, with
their bivalve shells and almost mussel-like appearance, are
fresh-water forms, though again a great variety are strictly
confined to the sea. Unfortunately the records in this case
are more incomplete than for the Cladocera, but there seems
little reason to doubt that the district would furnish a considerable collection to anyone expending time and trouble on
the matter. All the forms at present recorded appear to
belong to the single genus *Cypris*, if we except a note by
Jenyns which seems to refer to the form now known as *Noto-
dromus monachus*. *Cypris fuscata* of a large size has been
taken in great numbers on Sheep's Green, and we may also
mention *C. virens* and *C. pubera*, the latter from Bottisham
Fen.

THE FLORA OF THE CAMBRIDGE DISTRICT.

By A. Wallis, B.A., King's College.

The object of this short account of the Flora of the Cambridge district is not to give a list of localities, but rather to direct the attention of the reader to the vegetation as a whole. In the first portion of the paper will be found a short analysis of our flora and an attempt to explain the present distribution, that is, the Cambridge flora in relation to that of the rest of England.

The remainder of the paper is occupied with an account of the plant associations found in the district or the local flora.

By the Cambridge district is meant that portion of the county which is within easy reach of the town either by train or bicycle—that is to say, it consists of the whole of South Cambridgeshire, a small strip of North Hertfordshire, a corner of Essex and the western border of Suffolk.

Watson classified British plants according to their distribution in the Island. Those that were generally distributed he called British, those that were abundant in the south and tended to disappear towards the north he called English. He found that there were certain plants restricted to the East of England, to which he gave the name Germanic, not thereby implying that they had their origin in Germany, but using the word simply as a name to signify that they were plants found on that side of England which bordered on the German Ocean. In the same way those that had their centre of distribution in the west, he called Atlantic.

Combinations of these, such as Germanic English, would signify that the plant had its greatest density of distribution in the east but spread into some of the midland and south midland counties and even to some of the more easterly of the western counties.

If we apply this classification to the plants of the Cambridge district we have the following result.

English	173	Scottish	5
Eng. British	65	Scot. British	7
Eng. German	45	Scot. Intermediate	1
Eng. Intermediate	3		—
Eng. Atlantic	1		13
Eng. Local	4	Intermediate	4
	—	Int. English	1
	291	Int. Local	1
British	370		—
Brit. English	117		6
Brit. Scotch	7	Atlantic	[1][1]
Brit. Intermediate	3	Atl. English	3
Brit. Germanic	1		—
	—		4
	498		
Germanic	49	Total species 902.	
Germ. English	24		
Germ. Local	12		
Germ. British	5		
	—		
	90		

[1] *Erodium moschatum,* a doubtful native in West Suffolk, but possibly still to be found between Burwell and Upware on the grassy droves.

For comparison the flora of Berkshire is given.

English	183	Scottish	1
Eng. British	67	Scot. British	9
Eng. Intermediate	3	Scot. Local	1
Eng. Germanic	47		—
Eng. Atlantic	1		11
Eng. Local	3	Intermediate	2
	304	Int. British	1
		Int. English	1
British	379	Int. Local	1
Brit. English	111	Int. Highland	1
Brit. Germanic	1		—
Brit. Scotch	10		6
Brit. Atlantic	1	Atlantic	2
Brit. Intermediate	2	Atl. Local	1
Brit. Highland	3	Atl. British	2
	507	Atl. English	3
Germanic	20		—
Germ. English	23		8
Germ. British	4	Total species 886.	
Germ. Local	3		
	50		

The most noticeable feature of the Cambridge analysis is the comparatively large number of Germanic plants, 90 out of a flora of 902 species, or practically 10 %. In the Berkshire list it will be seen that the Germanic plants are about half as many as in the Cambridge list, they form but 5·6 % of the total flora; in Oxfordshire the percentage is the same; in Hampshire it falls slightly to 5·5, and in Surrey further to the east it rises to 6·8.

Further figures might be quoted, but sufficient have been given to show that there is a gradual shading off of Germanic plants towards the west of England.

It would be well to examine the Germanic plants with a view to obtaining further light on their distribution.

They are not a simple group but fall readily into three:

(i) Water or Fen Plants, such as:—*Stratiotes, Liparis, Senecio paludosus*[1] and *palustris*[1], *Sonchus palustris*[1] and *Damasonium.*

(ii) Wood Plants, such as:—*Primula elatior, Melampyrum cristatum, Hypopitys Monotropa, Ruscus* and *Convallaria.*

(iii) Plants growing in exposed dry situations, such as:— *Silene conica* and *Otites, Medicago falcata* and *minima, Trifolium ochroleucon, Seseli Libanotis, Veronica triphyllos, verna* and *spicata, Aceras anthropophora, Orchis ustulata, Allium oleraceum* and *Apera Spica-venti.*

The restricted distribution of the first group can be understood when their habitat is taken into consideration. They are fen plants and the fens are characteristic of the east of England.

This can be expressed numerically. If the census numbers, as given in the 9th Edition of the *London Catalogue,* are taken and the fen Germanic flora compared with the dry Germanic flora, it is then found that the average number of divisions of the dry Germanic flora is 18, whilst that of the fen is only 13.

The second group is not so satisfactory, and consists of but few species. Two, *Primula elatior* and *Melampyrum cristatum,* seem to be almost limited to woods upon the boulder clay overlying chalk, and as such their distribution is understandable. *Ruscus aculeatus* requires a hot sun and light soil to perfect its fruit. *Convallaria* is probably limited in its range owing to the same cause. There is no one factor that can be said to control the distribution of the shade Germanic plants.

The group of plants inhabiting open ground, either chalk or sand, is a large one. Their restricted distribution is not merely a question of chalk or no chalk, sand or no sand, for there are extensive sandy commons in Surrey, Berkshire and

---

[1] Probably now extinct in the district.

Hampshire, and chalk downs abound, yet in these counties
there are about half as many of the dry Germanic group as are
found in the Cambridge district, and further to the west there
are still fewer. To the north in Lincolnshire and Yorkshire
there is also a decrease. Sands and Limestones are found
all over England, yet this group of plants in its fullest development is restricted to the South Eastern portion of the island.
This restricted distribution cannot therefore be entirely due
to soil; climate must play its part.

The climate of eastern England and of the Cambridge
district in particular is drier and more sunny than that of
any other portion of England.

Combined with this we have a chalky soil overlaid in
many places with sands and gravels, the whole forming a most
perfect filter-bed which absorbs water exceedingly rapidly;
nor should it be forgotten that on this soil natural woods
are extremely scarce ; what trees there are have been largely
planted. There is but little to hold the water in the soil
and in consequence evaporation is rapid; these conditions all
working together tend to produce a state of affairs which
would encourage xerophytic plant life, and the great hairyness
of *Verbascum pulverulentum, Filago apiculata* and *spathulata*,
the depressed habit of growth combined with a long tap-root
of *Linum perenne* and *Artemisia campestre* and the very
small leaf-surface of *Iberis amara* and *Silene conica* are
examples of this.

The dry Germanic plants might confidently be expected to
show modifications which enable them to exist in the east
whilst elsewhere in England they are choked out by a more
water-loving vegetation.

Growing with them should be found other species, possibly
less specialised and so fitted to resist greater variations in
moisture and situation, yet all showing some xerophytic modifications, being less restricted in their distribution. These do
not come under Watson's heading of Germanic, although with
the true dry Germanic plants they may be regarded as forming
a loose association.

In East Anglia they would tend to be abundant, elsewhere they would be few in numbers, although widely scattered through many districts.

Such for example would be *Cerastium arvense* and *Avena pratensis.* The distribution of the former may be quoted to illustrate this. Its census number in the *London Catalogue,* 9th Edition, is 69. Round Cambridge in dry chalk or sand country it is exceedingly abundant, often forming quite a feature in the landscape, yet in Berkshire, Oxfordshire and Hampshire it is not common though scattered, and in Somerset it is confined to two localities, one near Bristol, the other on the Mendips.

Although it is easy to suppose xerophytic modifications in this association, it is not so easy to prove it. Some plants, like *Verbascum pulverulentum* with its dense downy covering and *Avena pratensis* with its jointed leaves, are obvious, others, like *Anthyllis vulneraria* which rolls its leaflets upwards over numerous sunken stomata, and yet possesses a few ordinary stomata on the under surface, are doubtful.

As to the great mass of our flora it is impossible to make any statement, too often it is a question of position of stomata, thickness of cuticle, &c. which necessitates microscopical observation; this has not been done, nor is there time to do it before this paper goes to the press.

Hairyness has therefore been taken as a test of xerophytism, for in any flora the fact whether a plant is glabrous or not can at once be noted without going to the actual specimen.

All plants mentioned as being hairy or downy in Bentham and Hooker's *Students' Flora* were marked off, and those rejected which grew upon damp heavy soils or in water. It may be pointed out here that the term "Xerophytes" includes three groups of plants,—salt-loving plants or Halophytes, plants living in dry well-drained soil, and plants living in badly drained soil where evaporation being rapid upon the surface so cools the underlying soil or water that absorption is very slow. Plants belonging to the second class were taken as being alone suited to a comparison dealing solely with a dry county flora. It was then found that 17% of the English flora consisted of hairy plants inhabiting light soils, 21% of the Cambridge flora were hairy, and for comparison the flora of Berkshire was taken as being a county where there are abundant chalk downs and sandy places; this flora showed 19% hairy; this slight difference would seem to represent the different effect of the climate in the two districts Cambridge and Berkshire.

If this test of hairyness be applied to the Cambridge dry Germanic plants alone, it is then found that 50% are hairy; this seems to show that the Germanic species are the most specialised members of this dry association.

There are numerous objections to this test; plants probably not xerophytic have been included, and the mere fact of taking county boundaries at all is artificial, including as they do wet and dry land in different proportions; yet here again one is held down by lack of data and lack of time. To make what would be a fair comparison, an estimate of the number of hairy plants to an acre of dry land as similar as possible in composition and position in a series of districts stretching north, south, and west would be necessary. Under the circumstances this is impossible.

In spite of these objections to the comparison given above it has been thought well to let it stand as probably representing, very imperfectly, the true condition of vegetation in this and the other districts mentioned.

What has been said of the dry Germanic group applies equally to the other two groups mentioned. The shade Germanics and the fen Germanics have growing with them other species of wider distribution.

In the case of the woodland plants it was found that their restricted range was due to a diversity of causes. There was no one influence, such as climate, which affected the whole group. The relation between the shade Germanics and the other shade plants with which they live differs from that

which exists between the dry Germanics and their associates.
The former are one of a great association which depends upon
subdued light and moisture, but they are not the specialised
members of that association as are the dry Germanics of
theirs. In this again they differ from the fen plants, which
agree with the dry Germanics.

To sum up then, we have in the Germanic flora three
groups—water, wood and dry soil plants. The first and last
of these with species of wider distribution combine to form
associations of which they are the most specialised members.
The second, the wood Germanics, restricted in their distribution owing to various causes, form but a portion of a wood
association.

PART II.

If the map at the commencement of this paper be referred
to, it will be seen that the Cambridge district has been
divided into areas coloured to represent the different plant
associations.

1. The fens indicated in light blue.
2. An area coloured pink, yellow and brown, a xerophytic
flora.
3. Meadow lands, blue, following the course of the rivers,
a grass association.
4. A green portion to the south-east of Cambridge with
an outlier to the west. This may be called the shade
association. Where the country is wild this has been indicated by an intenser shade, in the case of the fens bright
blue and of the upland flora bright pink.

It will be noticed that the floral areas numbered 1, 2,
and 4, have already been touched upon in the first portion of
the paper, the third, the meadow association, has no Germanic
representatives and so has hitherto not been referred to.

Before commencing the descriptions of the association, it
is necessary to point out, with regard to this map, that owing

to a variety of reasons, chief among these being an insufficiency of time to go over all the ground, the boundaries of the different areas are too often only approximately correct[1].

1. *The Fen.*

The great alluvial flats which are known as the fens are for the most part under cultivation ; here and there a rough meadow may be seen, still more rarely a piece of unreclaimed sedge fen. Fields of potato, bean, and corn with their attendant weeds of cultivation are what to-day compose the main flora of the fens.

Yet even now these expanses differ very considerably from the high land, the place of hedges is taken by ditches fringed with *Phragmites*, which spreads on either side in the black peaty soil and stands green among the yellow corn.

Where, as at Wicken, a portion of untouched fen remains, it is possible to see how marked originally must have been the distinction between the fens and the dry land.

Now that roads have been built and repaired often with chalk, plants foreign to the district are introduced. These flourish only along the roadsides, upon heaps of road metal, and upon the raised road dykes themselves. Such plants as *Anthriscus vulgaris*, *Reseda lutea* and *Festuca rigida* are not uncommon, whilst principally upon the river dykes are *Geraneum molle* and *Caucalis nodosa*. Besides these colonists are the weeds of cultivation in the fields, annuals which come and go with the crops.

The two great distinguishing features which separate the fen from the dry land are the absence from the fen of xerophytic plants such as *Hieracium Pilosella* and the presence of ditches, which take the place of hedges. This change can be well seen on the western side of the Wicken

---

[1] The map published with this paper is far from complete; it only serves to indicate the broader features and, it is hoped, will lead others to take an interest in the vegetation of Cambridge and so to complete, what has hardly been begun, a true Botanical Survey of the district.

peninsula. On the land are hedges with their dependent plants *Nepeta Glechoma, Anthriscus sylvestris, Galium Aparine,* and *Bryonia dioica.* On reaching the fen the hedges cease abruptly, the line of field division being continued by ditches fringed with *Phragmites.*

It is this presence or absence of ditches and hedges which forms the readiest method of determining the boundaries of the fen.

But few undrained portions of fen now remain. Four isolated portions are found occupying the valleys; one in the valley of the Cam near Hauxton (Dernford), another near the source of Quy water close to Fulbourn station, a third, the most important, at Chippenham and two small pieces in the valley of the Lark near Mildenhall. Besides these valley fens, which have been included with the real fens for convenience sake, there are small portions at Quy, beside the Bottisham Lode, Stalode Wash between Burnt Fen and Lakenheath stations, and the most important of all, Wicken Fen. This last is worthy of some description.

Wicken Sedge Fen which, with St Edmund's Fen, consists of nearly a square mile of country, lies to the south of the curved uplands of Wicken, north of Burwell Lode and east of the Cam, and is distant about eleven miles from Cambridge. Originally the fen must have been bounded on two sides by downs and must have stretched away to the south for some considerable distance over the country which is now known as Adventurer's Fen. Now it lies at some little elevation above the surrounding country, being bounded by broad ditches retained by dykes.

This elevation is probably due to the combined effects of turf cutting in the surrounding fens, formation of turf in the fen itself and shrinkage due to evaporation in its drained surroundings.

In dealing with this portion of the Cambridge district it is best not to regard the fen as distinct from the high land but to consider both together, tracing the transference of dry plant life upon the raised peninsula to the water and bog flora of the fen.

This is best seen upon the western side ; in former times the transference must have been gradual, now, owing to the artificial condition of the fen, it is irregular and abrupt. In this portion two zones of vegetation can be detected.

The dry land is for the most part cultivated; only by the Upware marl pits does any trace remain of its original flora. This flora though small is in very striking contrast to that of the fen. The following is a list of some of the more typical plants given in order of their importance.

> *Plantago media.*
> *Trisetum pratense.*
> *Phleum pratense* var. *nodosum.*
> *Cynosurus cristatus.*
> *Achillea millefolium.*
> *Medicago lupulina* (downy form).
> *Lotus corniculatus.*
> *Galium verum.*
> *Cerastium arvense.*
> *Festuca rigida* ⎫
> *Caucalis nodosa* ⎬ on broken ground.
> *Peucedanum sativum* ⎭

Below the uplands, in the first zone, consisting of rough meadows intersected with ditches, are found *Ranunculus Lingua, Sparganium minimum* and *Carex Pseudo-cyperus* in the ditches. *Holcus lanatus, Cnicus arvensis, Senecio Jacobaea,* and *Potentilla anserina* infest the meadows of this zone.

The second zone is still rougher; it is studded in places with old hawthorn bushes, which shelter a few shade plants, but the greater portion is bare. The general effect is that of a moor, but the flora has been so changed by the introduction of foreign soil, that a list of plants does not convey this impression. *Viola hirta* and *V. ericetorum* abound; *Arabis hirsuta* is common; *Linum catharticum, Primula veris* and *Briza media* are found.

The bank which retains the ditch surrounding the fen frequently undergoes repair; gault and chalky marl are used in its construction, and on it at first are found plants which are not native to the district. *Diplotaxis muralis* first springs up, and is after a few years replaced by *Cnicus arvensis* and poor grasses.

The fen itself appears during most seasons of the year as a brown waste dotted with bushes which are apparently increasing in number; in some places, especially along the northern edge of the fen, they form dense masses of cover, excluding all other vegetation. Elsewhere, when not entirely absent, they are scattered and usually smaller in size.

The great mass of vegetation however is composed of *Cladium jamaicense,* Crantz. This may be said to be the dominant species, but everywhere *Molinia varia* is found which competes vigorously for first place. Round the edges, where the fen is damper, *Phragmites* completely ousts both the above-mentioned species and reigns supreme.

The following two lists give some of the more important species inhabiting the dry and damp portions of the fen.

Dry.	*Wet.*
Cladium jamaicense,	*Phragmites communis.*
Molinia varia.	*Thalictrum flavum.*
Rhamnus catharticus.	*Lathyrus palustris.*
R. frangula.	*Convolvulus sepium.*
Salix, Spp.	*Caltha palustris.*
Cnicus pratensis.	*Lastraea Thelypteris.*
Peucedanum palustre.	*Jungermannia,* Sp.
Calamagrostis epigeios.	
Potentilla sylvestris.	

The "wet" plants seem to be a very natural group. The climbers, using the tall plants as a support and matting the whole together, form a dense mass of vegetation which shelters and shades the Lastraea and Jungermannia.

It should be noted, however, that what has been called the wet group is only to be found as a group round the skirting ditches and is best developed on the western edge of the fen. In these places where the ground is subject to frequent inundation the Phragmites ousts the Cladium; elsewhere, where flooding is of rarer occurrence, the Cladium, being better able to resist the drought, drives out the Phragmites. While it is not usual to find those species mentioned under the dry group encroaching upon the wet belt, yet the "wet plants" may frequently be found among the plants of the dry group.

It seems probable that the two transition zones of vegetation already mentioned, depending as they do upon water-supply, will be able to be traced round the fens. They occur at Chippenham Fen, and from near Mildenhall to Lakenheath in Suffolk, in some places better, in other places not so well marked, as on the western side of Wicken.

Although their characters are roughly the same in these places, rough meadows and what has been called moor, their plant associations vary in the first place with the soil, and secondly with human influence. At Chippenham, for instance, the chalk uplands masked with loose sand are almost entirely under cultivation, but their lost flora can be reconstructed from similar hill-sides in Suffolk. The meadow-zone yields such chalk plants as *Blackstonia perfoliata* and *Gentiana campestris*, while the moor is a Rush and Carex association. These two zones pass one into the other, the whole forming a gentle slope ending in the fen which resembles Wicken in its abundant growth of *Cladium* and *Molinia*.

Along the Suffolk border the slope is more gradual than at either Wicken or Chippenham. The upland flora is for the most part undisturbed, but will be referred to later; the transition zones are much entangled. The dominant plant in the first is *Saxifraga granulata*. This either covers the field with a sheet of white or occurs in local patches, according as the land is in the meadow zone, partially in it, or in the moor zone. The occurrence of the Saxifrage in such patches produces some extraordinary mosaic effects of colour in early summer. The ridges which are very irregular are a pure white, the hollows a deep green with grass and rushes.

In some places, as near Holywell, depressions in the sandy common are lined about a yard broad with white Saxifrage; this ribbon follows the windings of the banks, occasionally spreading over some low spur (first zone); below comes the deep green of grass (*Agrostis*), Rush and *Potentilla palustris* (second zone). Along this Suffolk border however the fen has disappeared and is represented by bean and potato fields or corn fields.

In Dernford Fen, south of Cambridge, the same may be observed.

2. *The Dry land.*

The lands lying above the level of the fens, either as isolated spots, islands, or as part of the system of hills to the south, are coloured pink, yellow and brown in the map. The whole represents a slightly xerophytic group of plants corresponding to the dry Germanic group and its associates already mentioned. This portion is characterised by light soils, chalk and gravel, and in West Suffolk by sand. Trees are rare and are generally planted in the form of wind-shields. Beech on the chalk and conifers and beech mixed on the sands of the Mildenhall district.

The planted trees alone, apart from other changes in the flora, would give a clue to the general division of this portion of the district.

In the west, Gorse and Bracken are rare, *Calluna Erica* probably not found, and *Carex arenaria* is absent; but these are characteristic of the eastern portion, whilst *Astragalus danicus, Linum perenne, Anemone Pulsatilla, Helianthemum chamaecystus, Trifolium ochroleucon, Polygala calcarea* are typical of the chalk downs, and *Carum Bulbocastanum, Fu-*

maria parviflora, Specularia hybrida of the arable land of the western district, and local or not found in the eastern.

This division has been marked on the map. The eastern portion is coloured yellow with a few brown patches, the western pink. To the east are found sands and gravels, covering chalk; to the west, to a large extent, chalk.

The large pink area, however, does not by any means represent a purely chalk flora; it is rather all those parts of the dry flora that remain after such obvious plant associations as the Carex and Bracken, and the Heather associations (to be referred to presently), have been separated from it. Other such groups might quite possibly have been found with a more careful survey of the county, but time has not permitted.

Everywhere over the western or pink area is the work of man apparent, the land is almost entirely under cultivation. A few semi-disused chalk-pits retain the remains of the ancient flora, which persists also upon the Fleam and Devil's Dykes and Roman road over the Gogmagog Hills. On these older works which have been undisturbed for centuries can be found *Anemone Pulsatilla* and *Senecio campestris*. The turf is close and springy and composed of grass and thyme. Where rabbits disturb the soil annuals appear; *Myosotis collina* and *arvensis*, afterwards followed by *Carduus nutans*. Here and there orchids are found, and on the Devil's Dyke *Thalictrum collinum* is abundant.

Before passing on to the eastern division it would be well to take the flora of one of the chalk pits, such as that at Haslingfield, as an example of the chalk flora. This pit is still used, but the greater portion is overgrown either with May bushes or grass; the former occupy the gullies in the sides, the latter the ridges. The flat bottom of the pit is cropped with sainfoin.

The vegetation as a whole may be said to show the effect of the soil. Only where the whitethorn has established itself is the vegetation rank; here *Arrhenathrum avenaceum* is abundant and *Galium Aparine* sprawls over the ground and climbs among the lower branches of the bushes. On the ridges *Festuca ovina* is dominant, *Leontodon hispidus* and *Hieracium Pilosella* abundant, and some grasses such as *Avena pratensis* and *Koeleria cristata* common in places. A count of plants on a square yard of rather bare soil gave

125 plants of *Festuca ovina.*
40   „   „ *Leontodon hispidus.*
8   „   „ *Hieracium Pilosella.*
3   „   „ *Gentiana campestris.*
2   „   „ *Polygala calcarea.*
2 seedling grasses *Avena pratensis.*

The ground had been recently disturbed and consisted of small angular pieces of chalk, favouring a strongly xerophytic flora. On a soil less recently disturbed some of the Leguminosae appear, notably a very hoary form of *Medicago lupulina* and *Trifolium ochroleucon.*

The eastern division unlike the rest of the Cambridge district is largely open country, man has hardly influenced it; here and there plough lands may be seen, and round Mildenhall a certain amount of tree planting has been attempted, but as a whole the soil is too poor to repay any labour spent upon it. Rabbits abound and these affect the flora to a very considerable extent (see p. 227).

The strata north and south of the river Lark differ slightly, and with them the flora. To the north over a great extent of country, from Mildenhall to Lakenheath and Thetford, two species of plants are dominant, *Carex arenaria* and *Pteris aquilina.* These do not occupy the same ground; there may be patches one hundred yards square of either, and it is not at all obvious which is driving out the other or, if they are in a state of equilibrium, what cause determines the distribution. Possibly the nearness of the chalky subsoil to the surface, by affecting the drainage, may determine this, but until more work is done on this question it is impossible to speak definitely. It has therefore appeared good to class the

Bracken and the Carex together, and to call this district where they are dominant the Bracken-Carex association. Not that they occupy all the ground; some portions of what is essentially the same country, especially round the Cambridgeshire border, are under cultivation and other portions have been ploughed up but have relapsed into waste.

In this northern portion Ling (*Calluna Erica*) only occurs here and there, and always in a secondary position, neither is Gorse very common. Both these plants seem to suffer considerably from rabbits. The comparative rarity of the *Calluna* seems due to the large amount of calcareous matter that the sandy soil contains, as the chalk is nowhere very far below the surface. If the surface soil is thick this calcareous matter, being less in amount, will not affect the flora to the same extent as on a shallow soil, where the plants are continually receiving fresh supplies of chalk. Hence Ling, being a calcifuge plant, should be expected upon deep surface soils. This it is not possible to determine without digging. As a rule however it is found in the depressions between low hills, and since the sandy covering would by denudation tend towards the valleys it is in these slight hollows that the greatest depth should be found.

South by the river Lark, where there are deposits of postglacial gravels overlying chalk, Ling is the dominant plant. Here upon Tuddenham heath, further to the south upon Kentford heath, and probably in one or two other scattered localities it is abundant. Between Tuddenham and Kentford heaths chalk covered with blown sand occurs, and here it is exceedingly rare. Carex and Bracken take its place, as they do in a narrow strip between the *Calluna* area and the rough meadows bordering the river. The river has worn away the greater portion of the gravel, the chalk is but a little distance below the surface and the Ling cannot grow.

This eastern district, presenting as it does in the main but two types of vegetation, Carex-Bracken and Heather, can yet when examined in more detail be broken up into several plant areas. One of the most striking of these is the grass-association. This does not occupy much ground, but there are patches scattered about, mostly round villages, so that it seems possible that it is one of the expressions of former cultivation. *Koeleria cristata*, *Festuca ovina* and *Galium verum* are the most prominent species, with here and there a plant of *Silene Otites*.

The whole is generally kept short by rabbits. Where they make their holes *Sedum acre* springs up to the exclusion of all else save of a few annuals. But this will be referred to later when the action of the rabbits is discussed.

Round the edge *Carex arenaria* is found encroaching upon these grass associations, so that it seems probable that they are one of the last stages in the reversion to heath of originally cultivated land.

Nor are intermediate stages wanting between this and crops.

In these the land still retains the original ridge and furrow formation of ploughed land, the vegetation is scattered and there is much bare sand. Some have been sown with grass and some have been left to revert naturally. In these last before competition becomes keen *Herniaria glabra* is exceedingly common; in others *Senecio Jacobaea* occupies the whole field, a mass of yellow, and were it not for the ravages of the Cinnabar moth caterpillar it would spread over the neighbouring county to the exclusion of all other vegetation. The *Senecio* suffers so much that in bad seasons but few plants ever set their seed; leaves, flowers and even the softer portions of the hard stems are eaten. Besides these remains of cultivation, there are other places where either the Carex or Bracken fails, which are either quite bare or are covered with a close crop of *Cladonia*. This lichen growth probably represents one of the stages in the reappearance of vegetation over deserts made by rabbits.

The rabbit affects the vegetation in two ways, firstly by eating, which in the main is destructive, secondly by burrowing, which is both destructive and reconstructive.

The grass suffers most destruction by eating, as can be seen where portions of land have been enclosed within netting; here grass rapidly springs up ousting the Carex. In the open, on the other hand, Carex, which is practically uneaten, drives out the grass. *Cladonia* and *Sedum acre* are also untouched and abound in places.

When burrowing is carried out on a large scale by many rabbits it produces what are practically deserts. The rabbits bore into a gently sloping hill-side, the soil falls down, a slight escarpment is made, and they bore again. This process, continually repeated, gives rise to considerable extents of loose sand, bounded on the upper side by a miniature cliff, full of burrows, on the lower side merging almost imperceptibly into the hill-side. The action of the wind upon the loose sand is such as by purely mechanical means to prohibit the growth of any vegetable life, but where stones offer any protection against the moving grains *Cladonia* will often be found. This may either cover in time the whole bare area or give way to *Festuca ovina* which in its turn gives way to Carex. Towards the lower edge the *Cladonia* increases considerably, with here and there a tuft of *Festuca ovina* and the straight lines of Carex shoots, until the normal growth of the undisturbed hill-side is reached.

So far the effect of the rabbit upon the vegetation has been principally destructive; there is however one class of plants—the annuals—which seem to depend for their very existence upon the isolated earths dotted over the heath. These heaps are at first bare, then when the hole is deserted, or but little used, annuals spring up. The following have been observed.

> *Erodium cicutarium.*
> *Veronica arvensis.*
> *V. verna.*
> *Myosotis collina.*
> *My. versicolor.*
> *Filago minima.*
> *Aira precox.*

This list could probably be greatly expanded; as it stands it is too small to generalise from. It is worth noting however that two, *Myosotis collina* and *versicolor*, are peculiarly fitted for animal distribution.

As the earth gets older, the annuals are replaced by perennials, but of a different sort from those of the surrounding vegetation. *Sedum acre* and *Cerastium arvense*, especially the former, abound. These seem to hold their ground well, particularly if the soil is loose and very dry, as is the case upon the hillocks which are so often chosen by the rabbits for their holes.

Here on this heath we have almost a primitive state of vegetation; man plays so small a part in the economy of nature that he can almost be neglected. Here we find certain species occupying great areas of country, undisturbed and unchecked by any agricultural operations.

Annuals are rare plants, what we now call corn-field annuals exceedingly rare. These depend for their very existence upon the constant and regular disturbance of the soil. Originally such plants, in those portions of the world where they had not invaded the cultivated land of primitive man, must have led a precarious existence upon land-slips, bare and crumbling river banks, but principally upon the earths of burrowing animals.

It is on the rabbit earths and on these alone that, in the wilder portions, annuals can exist. We see to-day the rabbit performing, in this quiet corner of England, his ancient *rôle* of agriculturist.

3. *The Meadows.*

The meadow association, which is marked on the map grey blue, occupies the river valleys of some of the streams round Cambridge. It is seen at its best where the rivers have cut through the chalk to the gault, and formed a flat valley bordered by low hills, which are for the most part cultivated.

These damp valley meadows are in such marked contrast

with the dry uplands that it seemed wise to recognise their
individuality by marking them upon the map of the Cambridge district. They have been by no means fully indicated;
a considerable portion of the river valleys that have been
coloured pink will, with more careful work, in all probability
be found to belong to this meadow association.

The dividing line between these meadows and the very
similar meadows bordering the rivers in the fens is by no
means so well marked as is the difference between the upland
vegetation and the meadows themselves. The grasses are
usually of better quality; *Deschampsia coespitosa* is rare,
whilst *Lolium perenne, Hordeum secalinum, Cynosurus cristatus, Alopecurus pratensis* and *Festuca elatior* are common.
Such plants as *Erophila vulgaris* D.C., *Arabis hirsuta, Carduus nutans, Senecio Jacobaea* and *Holcus lanatus*, typical
of the fen meadows, are either very rare or absent in the
valley meadows.

Originally the transference must have been very gradual,
as is shown by the isolated fens which have been already
briefly mentioned. But just as the fens under man's influence
took on the general character of cultivated land, and those
portions which now appear as meadow support upon the peaty
soil a crop of poor grasses and thistles, so on the other hand
the valleys which were but arms of the fen stretching inland,
having a clay soil, support naturally the finer grasses like
Lolium.

The flora of these meadows calls for but little attention;
in the ditches *Glyceria fluitans* and *Veronica Beccabunga*
are the two most common plants; in the meadows themselves
the hybrid between *Festuca elatior* and *Lolium perenne* may
occasionally be found.

In some of the less completely reclaimed spots, as at Cow
Fen and Dernford Fen, may be seen some of the remains of the
old flora, confined to ditches and small ponds in the former,
but in the latter in a small piece of damp land by the railway
embankment as well.

At Cow Fen are found *Stratiotes Aloides, Hydrocharis Morsus ranae*, and *Hottonia palustris*, plants whose presence here recalls similar situations in the fens themselves. And at Dernford Fen are

> *Molinia varia.*
> *Cladium jamaicense.*
> *Schoenus nigricans.*
> *Epipactis palustris.*
> *Orchis incarnata.*
> *Pinguicula vulgaris.*

Plants which show a true fen flora.

Surrounding this small fen are meadows showing, as has already been mentioned, the transference zones between highland and fen. At Cow Fen are meadows as featureless as those of any other portion of the Cam valley.

This meadow association is the result of the drainage of boggy river valleys. It exists owing to damp and yet well-drained soil and owes its origin and maintenance to man.

4. *The Woods.*

The south-eastern corner of the district, coloured green, together with the roughly circular portion to the west of Cambridge, represents what has been called the shade association. These two areas differ from the rest of the district in being well-wooded. Irregularly-shaped coverts are scattered over the country-side, between them lie meadows and occasional cultivated fields, in the western portion are found besides these considerable stretches of derelict land.

The woods are mainly composed of oaks, which nowhere reach any great size ; beneath them is a thick undergrowth of hazel. This is cut every few years, and with the inrush of light and air spring up sheets of Oxlip, *Primula elatior*. The majority of species seem to profit by the cutting of the undergrowth.

The following list is characteristic of these woods.

Anemone nemorosa.
Viola silvestris.
V. Riviniana.
Vicia sepium.
Primula elatior.
Melampyrum cristatum.
Orchis mascula.
Allium ursinum.
Ornithogalum umbellatum (rare).
Scilla festalis.
Paris quadrifolia.
Bromus ramosus.

The most noteworthy feature of this list is the absence of the common primrose and the presence of the Oxlip; otherwise the list, with the exception of *Melampyrum cristatum* and possibly *Ornithogalum umbellatum*, which is a doubtful native, might stand for any wood in the south of England.

The distribution of *Primula elatior* will be briefly referred to when the remaining features of these two districts have been described.

The meadow land is chiefly remarkable for its immense numbers of Cowslips, which are seen in their full beauty in May.

The derelict lands of the western portion are the direct outcome of the high price of wheat during the first half of last century, and the subsequent fall in price; much land which was before this time good pasture was ploughed up, fine old chalk downs were ruined, but no part of the Cambridge district seems to have suffered so severely as these clay lands. They are at present practically valueless, they will not grow grain at a profit, they have not been turned back into pasture, they have neither been planted with fruit trees nor oaks, they have simply been left; the result is a poor matting of Couch and Yorkshire Fog, thinly covering the clay, whilst hawthorn, ranging in size from a tiny seedling to a great thorny bush, threatens to turn the land into an impenetrable thicket.

Mr Miller Christy first worked out the distribution of the Oxlip, *Primula elatior,* in England[1]. It occupies two tracts of country, one large roughly dumb-bell-shaped portion lying in South Suffolk, North Essex and South-East Cambridgeshire and North Herts., and another a small circular tract principally in West Cambridgeshire. These tracts, it should be noticed, coincide with the shade association mentioned in this paper. The Oxlip is a semi-shade plant occupying open woods upon boulder clay. Although along the north-west border in Cambridgeshire, stretching from near Bury St Edmunds to Saffron Walden and beyond, its limit of distribution practically coincides with the limit of the clay, elsewhere it is found always growing within the bounds of this deposit.

Probably within these areas, certainly within the larger of the two, the Primrose, *Primula acaulis,* except where planted, is not found; the Cowslip, *P. veris,* however, as already mentioned, abounds. Round the outside of the Oxlip areas hybrids occur, both between *Primula acaulis* and *Primula elatior* and *Primula acaulis* and *Primula veris*; elsewhere these are not found, but *Primula elatior* and *Primula veris* occur, scattered over these districts, but are always very rare.

Taking Cambridge as a centre it is easy to trace the north-west limit of the larger Oxlip area. Using the Newmarket line which at Six-mile-bottom rises slowly till it reaches the top of the chalk at Dullingham, it is possible, striking south from either of these stations, to quickly gain a yet higher ridge, boulder clay overlying chalk, where at once the woods begin. Along this ridge of hills one can trace the hybrids from copse to copse ; at the end of the Devil's Dyke, near Newmarket to Dullingham, Brinkley, Balsham, and again across the lowland from Hadstock to Saffron Walden. Going, at any point along this boundary, further south-east the hybrid belt is passed, and copses full only of the pure Oxlip are found.

To the west of Cambridge the district is much smaller, and here the hybrids are more numerous.

[1] *Journ. Linnean Soc.* Vol. xxxiii. 1897, pp. 172—201.

There yet remains one group of plants, the estuarine flora, of which hitherto no mention has been made.

Originally a large portion of the fens must have been under the influence of the tide. Until the two great cuts, the old and new Bedford rivers, were made and Denver sluice erected, the tide flowed up the Cam to within ten miles of Cambridge. Now the salt water is diverted and flows up the New Bedford river, ordinarily beyond Welney, and at spring tides as far as Mepal. A fresh-water tide is felt however beyond Earith.

These great drains are very difficult of access, roads are scarce and convenient railway stations still scarcer. It is nevertheless comparatively simple to approach them at Earith, where is the junction with the Ouse, at Mepal, at Welney, and again by train at Downham Market, where they again join the river.

The estuarine plants are however not abundant and are, as would be expected, for the most part near Downham Market.

The following short list probably includes the greater number of those to be found.

> *Cochlearia anglica.*
> *Aster tripolium.*
> *Scirpus maritimus.*
> *Glyceria maritima.*

How far these extend up the New Bedford River has not been ascertained. They were with the exception of *Scirpus maritimus* noticed by Denver sluice. The Scirpus is probably to be found at Welney.

Traces of old tidal influence are very rare. *Apium graveolens* in its more northern localities may be due to this, and *Scirpus maritimus* at Littleport, where it still exists or did till comparatively recently, is another example of an old estuarine flora.

There are besides these two species two more, *Viola lutea* and *Bupleurum tenuissimum*; the first in West Suffolk, the second on the Isle of Ely. This Viola in the south of England is usually found near the sea, though in the north it is a mountain plant.

These, unlike the others, are not estuarine species; they inhabit dry banks above the reach of the tide. In their Cambridgeshire and Suffolk localities they may be the last remains of an old sea flora which has disappeared, owing to its inability to live under the changed conditions which ensued with the interruption of the flow of the tide.

In closing this short account of our local flora I wish to express my thanks to Mr R. P. Gregory of St John's College for the unsparing help he has rendered me in preparing this paper, and to Mr A. H. Evans of Clare College for information on the distribution of some species.

NOTES ON COAST FLORA OF HUNSTANTON.

The sea-coast flora of Hunstanton falls into four fairly distinct groups :

1. Mud plants.
2. Pebble bank plants.
3. Chalk plants.
4. Sandhill plants.

The distribution of these groups is governed by the configuration of the land.

At Heacham, a small river runs into the sea forming a low-lying muddy estuary. The tide rip working against the river has pilled up a small pebble bank which reaches its maximum development at Heacham and gradually thins away to the north.

The town of Hunstanton itself is built upon a gently sloping chalk hill which terminates in a small cliff varying from fifteen to thirty feet high and about three-quarters of a mile long. To the south are found the plants characteristic of the Mud and Pebble bank, to the north of this cliff occur the Sand hill plants.

The Mud Plants.

The distribution of the mud plants is governed by their liability to flooding by the tide. Some, like *Glaux maritima*, are found almost on the beach or in very low-lying situations where they are frequently covered by water.

Others, such as *Aster tripolium* and *Glyceria maritima*, occupy the sides of muddy creeks which fill with spring tides, but sometimes for days together remain nearly dry.

Suaeda maritima and *Salicornia herbacea* occupy flat ground fairly frequently covered by the sea. On rather higher ground is found another group of plants, and in little dry knolls rising above any but the highest spring tides is *Frankenia laevis.*

This distribution has been expressed briefly below.

Very wet places.	{ *Polygonum Raii* ? { *Glaux maritima.*
Ditches.	(*Aster tripolium.* { *Glyceria maritima.* (*Gl. distans.*
Flat muddy places.	(*Buda marina.* { *Suaeda maritima.* { *Salicornia herbacea.* { *Triglochin maritima.* { *Juncus maritimus.* (*J. Geradii.*
Intermediate.	*Statice Limonium.*
Rather drier places less frequently submerged.	(*Statice auriculaefolia.* { *S. rareflorum* ? { *Salicornia depressa.*} (*Suaeda fruticosa.* }
Intermediate.	*Statice reticulata.*
Dry knolls.	{ *Frankenia laevis.* { *Armeria vulgaris.*

Suaeda fruticosa although occupying, so far as moisture is concerned, the same area with the two *Statices* does not grow with them.

It prefers rather more stony soil on the borders of the pebble bank and is found along with *Salicornia depressa.*

The Pebble Bank.

The pebble bank flora consists of two groups—one comprising deep-rooted plants which by means of long tap-roots penetrate the loose dry stones and grit and reach soil and moisture below—the other a shallow-rooted group which inhabit the less dry and barren portions of the bank, or depend for their existence upon the transitory moisture of the soil, with no permanent deep-situate supply.

Botanical Map
of the
Cambridge District

Fen.

Unreclaimed Fen.

Meadow Plant
   Association.

Shade Plant
Association.

Dry Soil Plant
Association, mostly
showing the influence
of Chalky Soil.

ditto. Wild uncultiv-
   ated land.

Carex & Bracken
Association.

Heather Association.

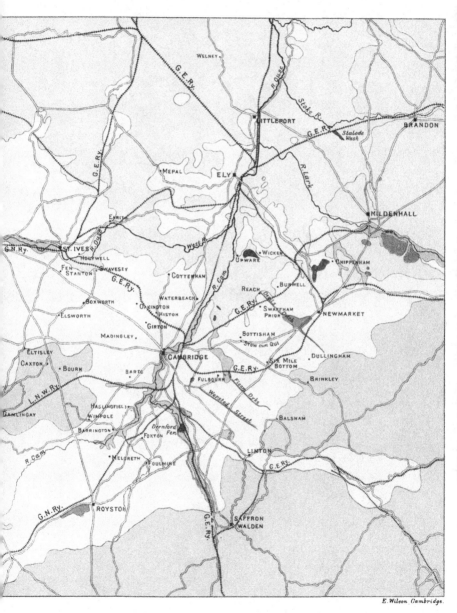

WELNEY

G.E.Ry.

LITTLEPORT

R. Ouse

Stoke R.

G.E.Ry.

BRANDON

G.E.Ry.

Stalode Wash

MEPAL

ELY

R. Lark

MILDENHALL

G.N.Ry.

EARITH

West R.

WICKEN

CHIPPENHAM

ST. IVES & OUSE

HOLYWELL

UPWARE

FEN STANTON

SWAVESEY

G.E.Ry.

COTTENHAM

R. Cam.

REACH

BURWELL

BOXWORTH

WATERBEACH

G.E.Ry.

SWAFFHAM PRIOR

NEWMARKET

ELSWORTH

OAKINGTON

HISTON

MADINGLEY

GIRTON

BOTTISHAM

STOW CUM QUI

DULLINGHAM

ELTISLEY

CAXTON

BOURN

CAMBRIDGE

BARTC

SIX MILE BOTTOM

G.E.Ry.

BRINKLEY

FULBOURN

Fleam Dyke

L.N.W.Ry.

HASLINGFIELD

WIMPOLE

Worsted Street

GAMLINGAY

BARRINGTON

Dernford Fen

FOXTON

BALSHAM

R. Cam.

MELDRETH

FOULMIRE

LINTON

G.E.Ry.

G.N.Ry.

ROYSTON

G.E.Ry.

SAFFRON WALDEN

E. Wilson Cambridge.

Deep-rooted.	{ *Glaucium flavum.* { *Eryngium maritimum.*
Intermediate.	*Silene maritima.*
Shallow-rooted.	{ *Arenaria peploides.* { *Trifolium scabrum.* { *T. striatum.* { *Phleum arenarium*[1].

The Cliff.

The cliff flora calls for hardly any notice. Wallflower, probably self-sown, is not uncommon on the cliff face; and fringing the edge is a great quantity of *Smyrnium olusatrum.*

The Sand Hills.

The sand hill flora consists of three sections.

1. Annuals which exist outside the area where the sand is held in check. *Cakile maritima* and *Salsola kali* are the most important.

2. The area where the movement of the sand by the wind is partially checked by the presence of some binding plants.

The principal binders are :

(1) *Ammophila arundinacea.*
(2) *Carex arenaria.*
(3) *Elymus arenarius.*

The first two are by far the most important. Protected by these plants are

Arenaria peploides.
Erodium cicutarium.
Volvulus Soldanella.
Phleum arenarium.

On the eastern side, protected from the wind, *Lactuca virosa* is abundant.

3. The region where no sand is in motion and the binders have given place to short turf—the links.

---

[1] *Ph. arenarium* rare on stones.

PREHISTORIC ARCHAEOLOGY OF CAMBRIDGESHIRE.

By W. L. H. DUCKWORTH, M.A., Fellow of Jesus College, University Lecturer in Physical Anthropology.

WHILE the prehistoric antiquities of Cambridgeshire are far from scanty in amount, yet it must be admitted that their character does not lead one to infer that the district was ever the centre of great activity or the scene of events of fundamental importance in determining the future history of our land. (The present account will not consider the history of the monastery of Ely, nor the events which render it so conspicuous from the historical standpoint, especially in the eleventh century.)

If one studies the nature of the locality, it becomes evident why prehistoric Cambridgeshire should present this particular character; for a county comprising so much fenland as that of Cambridge did until the seventeenth century, would naturally be little more than a refuge for those who by stress of circumstance were unable to subsist on the richer lands by which it was surrounded.

The creation of the county of Cambridgeshire is ascribed, though with small show of evidence, to Edward the elder son of Alfred the Great and to his sister. To quote a recent historian[1] of the county, Cambridgeshire is "a long strip of territory bounded on the east by the ancient East Anglian

---

[1] Rev. E. Conybeare, M.A., *History of Cambridgeshire.*

counties, and elsewhere by arbitrary lines the exact delimi-
tation of which was due, doubtless, to long-forgotten local
reasons of the tenth century. Its greatest length from north
to south is fifty miles, its greatest breadth about twenty-five,
either line of greatest dimension passing through the town of
Cambridge.

"The county is divided by Nature into two regions of
about equal size, but very different in character, the fenlands
in the north and the low chalk uplands in the south. The
latter, strictly speaking, form the actual shire of Cambridge,
the former having to a great extent a legally independent
recognition, under the name of the Isle of Ely....The region
thus privileged is now a vast alluvial plain, almost treeless,
intersected in every direction by a net-work of ditches, locally
called 'lodes,' from which the water is pumped by steam
power into the sluggish channels along which it makes its way
to the sea. The whole district is only kept dry by artificial
means, for it is well below sea-level, and even within living
memory was one vast morass, tenanted by innumerable wild-
fowl. To this day it remains sparsely inhabited, the few
towns and villages located on the almost imperceptible rises,
marking what were once islets amid the marsh. The elevated
ground on which Ely itself stands formed an island of greater
height and of respectable size, giving space for a whole group
of villages. It is still surrounded by water, various branches
of the Ouse stagnating around it.

"The Southern region consists of low chalky uplands,
through which the Cam and its sister stream, the Granta,
with their various tributaries, have eroded marvellously broad
valleys for such petty rivers. The former (also called the
Rhee) has its source in a lovely group of springs called
Ashwell, in Hertfordshire, just across the county boundary;
while the latter in like manner rises just outside the county
near Saffron Walden in Essex. The southern boundary of
their basin is formed by the escarpment of the great chalk
plateau which sweeps through England from the Yorkshire

Wolds to the coast of Devon, and is so conspicuous a feature in the geological map of our island, and which at this part of its course forms the watershed between the valleys of the Thames and the Ouse. The chalk itself is locally called 'clunch,' and is much used for building, some of the churches (e.g. Barrington) being entirely constructed of this material."

And yet the profusion of remains of high antiquity is not surprising, for Cambridgeshire is a district lying on several important lines of communication, thoroughfares of importance from the earliest period, either from their strategic value or their subservience to trade. For this reason the archaeology of the county is necessarily to a large extent associated with the great roads by which it has been traversed for so many centuries, and in general the character of the prehistoric remains is such as one would expect to find in the neighbourhood of trade routes and important military roads.

The present account therefore comprises in the first place some description of the ancient roads. Next in order it is convenient to consider the relics of early settlements in the shape of forts or earthworks. Following this, the consideration of the occurrence and the characters of the early implements, weapons, or other objects, will then claim attention. And finally a few notes will be added on the characters of the remains of prehistoric human beings which have so far come to light in the course of excavations of whatever kind.

I. *The Roads.* The accompanying map (no. I.) shows diagrammatically the course of six of the principal ancient roads, which will now be considered briefly.

The Via Devana. This name, like those of the other roads to be subsequently described, dates from about the 17th century and its exact origin is quite obscure. As a descriptive term, it is however quite appropriate, for the road in question affords communication between the ancient Roman stations of Chester and Colchester. Its general direction is still followed through Cambridge by Bridge Street and the Huntingdon Road, and traces of its use in Roman times were revealed in 1823, when a Roman causeway was displayed by excavations on the eastern side of Bridge Street. The Anglo-Saxon settlement discovered near Girton was not far distant from the line of the Via Devana, and near the same locality have been found Roman remains in the form of Samian ware, glass vessels, and two stones with inscriptions, one of which associates the Vth Legion with its erection. The Via Devana leads on this side of Cambridge to Godmanchester, and so *via* Leicester to Chester.

To the south-east traces of the road are less distinct: the road now called the Hills Road runs southwards and surmounts the Gogmagog Hills. About two miles outside Cambridge, a road leads off to the south-east, and following this, a great earthwork or dyke is found on the crest of the hill, broad enough for use as a road and running for miles in the same direction. This, the so-called Worsted Street, was claimed by the late Professor Babington as the continuation of the Via Devana, but more recent researches by Professor Hughes point to the Worsted Street as a defensive work, and thus another line must be sought for the Via Devana, and is most probably to be found in the actual line of the Hills Road which crosses the Gogmagog hills further west, on its way to Linton.

The Akerman Street. Another road of importance is that which crossed the Via Devana almost at right angles at Cambridge. The name is said to be derived from Akerman-chester, an ancient name for the station of Bath: the road leads from Brancaster on the Norfolk coast *via* Cirencester to Bath, and thus afforded an important means of communication between these strategic points. The point of intersection with the Via Devana was somewhere on the raised plateau above the Castle Hill, which reminds one that modern Cambridge is principally situated on the opposite bank of the river to that which was first inhabited, so that Castle End is the most ancient part of the town.

Traces of the Akerman Street may be sought for to the north-east and south-west of Cambridge. To the north-east, it is said to be represented in certain green tracts now lost in the fields in the neighbourhood of the "King's Hedges[1]" and the village of Histon. Roman coins have been found along this part of the road, which has been traced *via* Denny Abbey to Ely, Littleport, Denver, and Brancaster. Along the latter part of its course, not only coins, but bronze vessels, shields, copper mirrors, fibulae and beads have been found. In the opposite direction, the road is believed to have traversed the land on the western side of the Grange Road, and to have entered the Barton Road, passing near Comberton, where a Roman villa was discovered in 1842, to Orwell, crossing the Ermine Street near Wimpole, and so to Ampthill in Bedfordshire.

The two foregoing roads actually passed through the ancient settlement of Cambridge. We now turn to routes which are recognisable in other parts of the county.

The Icknield Way. The name is associated with that of the Iceni or Eceni, the predominant clan in this locality, at the time of the Roman invasion. If one traverses either the modern Hills Road over the Gogmagog hills, or follows the line of the Worsted Street over the same range, one meets at a distance of about six miles out from Cambridge, a great road, now fringed with telegraph poles, and leading from south-west to north-east towards Newmarket. Traceable from Royston to Norwich, it provided an access from the south and south-west to the eastern counties, only passing south of the fenlands round Cambridge, just as the Akerman Street passed to the northern side. The line of chalk hills traversed by the Icknield Way provided a safe and commodious route which avoided equally these fens on the north and the impenetrable forests (now destroyed) on the south. Besides the Roman antiquities (cinerary vessels and coin at Five Barrow Field

---

[1] Not far from site of an ancient camp: possibly Roman, or possibly used by William the Conqueror during his operations against Ely. But of the origin of the name "King's Hedges" nothing is known, nor is anything remarkable now visible on the spot.

near Royston) found near this road, evidence of earlier and British associations, in the form of hut-circles containing ashes and fragments of bronze similar to those found at Barrington[1], bear testimony to the great antiquity of this route.

But beyond all this, the Icknield Way is conspicuous as being crossed in Cambridgeshire by no less than three great earthworks or ramparts, viz. the Worsted Street already described, the Fleam Dyke, and the Devil's Ditch, of which mention will be made in the succeeding paragraph.

The Ermine Street. This, the great line of communication between London and York, the "Old North Road" of modern times, crosses the western end of Cambridgeshire on its way from Royston to Huntingdon; at the former town it crosses the Icknield Way, and at the latter it crosses the Ouse in company with the Via Devana. Its significance in the early history of Cambridgeshire is as yet unknown.

The Pedder Way is another road along the line of which numerous discoveries of archaeological interest have been made. Starting at Stratford-le-bow near London[2], and passing Woodford, Epping, Harlow, Bishop's Stortford, and Newport, it reached Great Chesterford, at about a mile beyond which it joined the Icknield Way, and they proceeded together at least as far as Worsted Lodge (on the Via Devana, i.e. the Worsted Street), and perhaps to Mutlow Hill Gap on the Balsham Dyke. Then passing towards the Beacon tumuli on Newmarket Heath (an extensive Anglo-Saxon burial ground was excavated by Lord Braybrooke near this portion of the Way), it seems to have cut through the Devil's Ditch at the "Running Gap" on its course to Exning, and here also many Saxon and Roman remains have been brought to light. From Exning it passed to Mildenhall, thence to Brandon, Ickborough, Swaffham and Castle Acre to Brancaster. As has just been mentioned, Roman and Saxon antiquities are numerous along this line. Thus at Great Chesterford an extensive Roman

---

[1] Cf. Conybeare, *op. cit.* p. 15.

[2] Babington, *Ancient Cambridgeshire*, p. 64.

station occurs (this is identified with Iceanum of the Romans). In the same neighbourhood a large refuse pit containing, among the rubbish, coins of Magnentius (A.D. 350—353), Valentinianus (A.D. 364—375), and Victorianus, with broken Roman pottery and also bronze articles, was investigated by Professor Hughes. At Mutlow Hill, bronze fibulae and numerous coins were found in the foundations of a circular building near a tumulus. The Beacon tumuli provided relics of a British cinerary vase and some bones and ashes. Finally Roman coins and urns were dug up at Exning (for the foregoing objects and their description, cf. Babington, *op. cit.* pp. 67, 68).

The Fen Road was a great causeway driven through the Fens, which it crossed from Huntingdon on the west to Norwich. It thus connected the Ermine Street (at Huntingdon) with the Akerman Street which it crossed at Denver. Traces only are now to be seen of the way, which was sixty feet in width and was raised some two or three feet above the surrounding lands. Near it at March there were found, in 1730, two urns, one containing bones and ashes, the other 300 silver coins "of all the Roman emperors from Vespasian to Constantine." Much Samian ware and sepulchral urns were found near Stoney, also near this way, and other Roman remains, such as pigs of lead, have occurred at localities supposed with more or less reasonableness to have been on the same line.

The foregoing notes must suffice for an indication of the chief ancient lines of communication through the county, and attention must next be turned to remains of ancient settlements or fortifications. The first of these to claim attention will be the well-known Castle Mound, a great tumulus situated on the north bank of the Cam and close to the former site of Cambridge Castle. This tumulus is not however to be regarded as of much greater antiquity than the 9th century (cf. J. W. Clark's *Guide to Cambridge*, in which a section is represented passing through the mound and neighbouring region). An outlying line of defence is still recognisable between the base of Castle Mound and the river bank, in the form of a raised terrace now in the garden of the Lodge of Magdalene College: further traces of the same rampart have been noted further west in the neighbourhood of what is now Northampton Street.

The Worsted Street has already been mentioned: if it be regarded as a defensive work, it would seem that its object was to provide protection against advances from the west, as the vallum is on this aspect of the rampart. As already mentioned, it is traceable from near the crest of the Gogmagog hills for several miles, reaching beyond the county border and ending indefinitely among cultivated fields. Near its Cambridge extremity and along the crest of the same Gogmagog range of hills there has been discovered a trench of considerable extent running along the line of the hills; though filled up with débris containing potsherds and fragments of cinerary urns from a Romano-British settlement, yet the trench has been recognisable as such within the memory of persons still alive, and the various portions thus visible went by the name of the "War-Ditches." If associated with the other great ramparts, it would be a work of pre-Roman antiquity: but nothing definitely proves this. On the other hand, remains of a Romano-British settlement are locally plentiful, and there is some probability that the ditch and bank originated with a Roman encampment. It is to be noticed that numerous skeletons have been found in the line of the filled-in trench, and that these interments are most likely of post-Roman date.

Vandlebury is the name given to a circular camp situated about a mile to the south-east of the "War-Ditches." It was surrounded by three ramparts, between which two ditches intervened: by the late Professor Babington it was regarded as a British stronghold occupied subsequently by the Romans as was indicated by the discovery of coins.

If the Worsted Lodge Dyke (Worsted Street) be traversed to the Icknield Way and this in turn followed for a mile to the left (i.e. north-eastwards) the great Fleam Dyke will be met with crossing the line of the Way almost at right angles. This and the Devil's Ditch are so similar to one another that they may be described together. Both Ditches are furnished with a mound or rampart on their eastern side; so that, if defensive works, their makers must have anticipated attacks from the west. Both run from the fenland in the neighbourhood of the Cam, and after crossing the open country between fen and forest, end in the latter, the Devil's Dyke crossing Newmarket Heath from the fens at Reche to the woodlands near Wood Ditton (Ditch-town); while the Fleam Dyke is traceable from near Fen Ditton (Ditch-town) to the wooded country near Balsham. But there are some differences to be noted between these two great Dykes: for the Devil's Dyke is straight throughout, while the Fleam Dyke is sinuous in correspondence with the requirements of the ground; and whereas the former is continuous throughout its length, the Fleam Dyke is interrupted at Wilbraham Fen (which rendered a special dyke unnecessary), and again shortly before its termination at Balsham. To the foregoing must be added mention of two other ditches, though these are of much smaller dimensions. The Pampisford Ditch, about a mile and three-quarters in length, is to be seen at Brent-Ditch End at Pampisford (eight miles from Cambridge) and it crosses the Icknield Way just as the three other ditches do; there is some uncertainty as to which side of the ditch was occupied by the rampart. Further to the south, and also crossing the Icknield Way, is the Brand or Heydon Ditch, two miles in length, with ramparts on the eastern side.

It is not possible to enter into a detailed description of these remarkable remains, but it must suffice to insist upon their impressive character, especially in the case of the larger examples, consisting as they do of the combination of bank and ditch traceable for miles across the county. Nor is it possible to discuss their significance. From the Heydon Ditch comes some evidence to shew that the Icknield Way existed before the trenches were cut. Though popularly supposed to be military works, it is possible that they may have equally well served the purpose of preventing cattle-lifting in ancient times. Professor Ridgeway supposes that they are referred to by Tacitus in describing the suppression of a rebellion of the Iceni and other East Anglian tribes by Ostorius (in A.D. 50). And finally as regards the associated remains; these consist principally of Roman coins (Devil's Dyke, Reche, Burwell Fen; Heydon Ditch, Heydon), of Roman vessels (Devil's Dyke, Bottisham and Burwell Fen), of bronze fibulae (Bottisham), and finally bronze and iron implements (at Heydon). Tumuli were once visible near the Fleam Dyke at Mutlow.

Among other instances of ancient settlements there may be specially noticed the important tumuli known as the Chronicle Hills near Triplow. Later settlements are those of Barrington, where remains of both British and Anglo-Saxon villages have been found. Nearer Cambridge itself are the sites at Comberton (Roman villa), and Grantchester, at which latter place are ancient earthworks commonly ascribed to the Roman period, while other localities have been already mentioned incidentally in connection with the various discoveries made from time to time, small settlements of the Romano-British period being particularly numerous.

Turning now to the general consideration of the flint implements which have been brought to light, we find that these objects carry us back, as was to be expected, to more early periods than any suggested by the antiquities hitherto enumerated. For the occurrence of Palaeoliths denotes the presence of man in a phase of cultural evolution far anterior to that attained by the British inhabitants of the county. It must however be at once stated that while Cambridgeshire itself is not fertile as yielding examples of ancient flint implements, yet a step over the county border into Suffolk brings one to perhaps the most important factory of these objects in the whole of Europe. To this day the trade in gun-flints persists at Brandon as a souvenir of the once far more extensive production of the various objects to which the names of axe, scraper, borer &c., are applied. From the borders of Suffolk then, especially from Mildenhall, where the excavations at Grimes' Graves (see Greenwall and Rolleston, *British Barrows*) still shew the method of flint extraction from subterranean workings in the chalk, from Lakenheath in the same neighbourhood, from the flint-strewn country between these places and Newmarket, comes a cloud of witnesses in the form of flint implements of palaeolithic age. From Cambridgeshire itself the number is comparatively small, nevertheless palaeoliths have been dug up in Manea Fen and in Burnt Fen[1].

Coming to the Neolithic objects, the same general remark as to their *provenance* holds good, again with the exception that Manea Fen and Chesterton have yielded good examples of worked flints of various sizes, those from Manea Fen being distinguished by their rich red-brown colour. But if we comprise Mildenhall and Lakenheath within our province, we may note a profusion of forms and extraordinary variety of objects of fine execution, and the Cambridge Antiquarian Museum can shew most excellent and complete series of axe-heads, knife-flakes, borers, scrapers and chisels, down to the tiny splinters which were perhaps used for bone-carving or even tattooing; special notice should be given to the wonderful collection of flint arrow-heads comprising flints of every shade of colour, and of at least four well differentiated forms (leaf-shaped, barbed, barbed and tanged, and triangular). Transitional forms between the chipped and the polished implement are also exhibited.

The greater number of implements recovered from the peat are of polished stone. A notable example is the implement found embedded in the forehead of an Urus (the great extinct ox of pre-Roman Britain), and now exhibited in the Geological Museum of the University.

The material (quartzite) must have been brought hither

[1] *Antiquarian Report*, 1899, p. 6.

from great distances, and not improbably the objects themselves were articles of barter and commerce, and this may not inconceivably have had a Scandinavian origin. A few instances of perforated polished axe-heads occur.

Objects of later date are the bronze swords and spearheads which have occurred at Barrington, and are now exhibited in the Antiquarian Museum; these with various objects described as personal ornaments have been obtained from settlements of Anglo-Saxon age at Barrington (cf. the Walter Foster Collection), and near St John's College. The two bronze shields found at Coveney Fen in 1846 are probably of earlier date and of British origin. In the same connection, mention is to be made of the remarkably rich collection of cinerary urns of the British period which are so well exhibited in the cases of the Museum, and the unique series of pottery of Roman and Saxon antiquity which have been exhaustively described by Messrs Jenkinson and von Hügel. As has been so often incidentally mentioned, finds of Roman ware and of the allied Romano-British pottery constitute a very large part, if not the majority, of the total of discoveries. At Horningsea, five miles below Cambridge, on the Cam, are visible traces of a Romano-British settlement particularly productive of the coarser kinds of ware, and such settlements are numerous throughout the county.

From the objects we now turn to the makers, and consider the evidence available as to the characters of the early populations of Cambridgeshire. It may be stated at the outset that there are no human remains possessing any evident title to be attributed to the palaeolithic period, and indeed the same may be said as regards the neolithic period as well, unless we regard the long-headed, short-statured British population as lingering representatives of the neolithic man of the locality. The custom of cremation has further reduced the number of examples to a comparatively small figure. Besides this it is remarkable that very few remains have been collected besides the skull: the long bones are usually not preserved with such

care by those who discover them, and for this reason there is
wanting most of the evidence upon which an estimate of the
stature and development in physique of the ancient inhabitants
of Cambridgeshire might be based.   Of skulls however a fair
number is available.   A general review of these shews most
distinctly that from the earliest period represented, whether
from the tumuli in association with the long-barrow mode of
interment, from settlements of Roman date, or from Anglo-
Saxon, or Romano-British cemeteries, down to mediaeval and
even later and modern times, the prevailing skull form has
been that of an elongated ovoid, described technically as
dolicho-cephalic.   The statement must be slightly qualified
by adding that the degree of narrowness is not extreme, that
the long skulls border on the province of those of mean proportions (i.e. neither very long nor very broad) which indeed
also actually occur.   But short broad skulls are distinctly rare,
and are not infrequently associated with post-Norman remains,
which suggests that the Norman Conquest may have been
accompanied by an influx of round-headed foreigners in this
district.   Such an incursion would be only one of many to
which it is practically certain Cambridgeshire in common with
the rest of East Anglia has been subjected from periods much
earlier than the 4th and 5th centuries when immigrations
occurred on a large scale.   Some of the immigrants may have
been round-headed Danes, but the majority were Scandi-
navians, Angles, and certain Belgic tribes, whose skull-form
was characteristically narrow, and who blended with the native
population with the resulting skull-form that is seen to charac-
terise the Romano-British of our collections.   It is practically
impossible to attribute any special specimens to particular
tribes or clans: thus we remain in complete ignorance of the
head-form prevailing among the Iceni and among the Girvii,
the principal clan of fen-dwellers.   At Brandon specimens
occur which have been identified by Dr Myers with a type
well-known on the continent as the "Batavian."

The average stature of such individuals as have been discovered in coprolite diggings, in the excavation of tumuli &c., is what would now be described as below the mean, though taller individuals are by no means absent. Physical development was not very different from that of the modern East Anglian peasant, though perhaps the evidence of the excavations made in search of flint at Grimes' Graves, and the proportions of the primitive picks of stag's horn used by the early workers there, points to the occurrence of individuals, or even to a tribe of persons, of very slight stature, and almost pygmy proportions. Against this we have plenty of examples of stout massive bones, as a particular case of which, the skeleton found associated with late Roman pottery at Horningsea may be cited, or the fine male skeleton discovered in the excavation of the War-Ditches at Fulbourn (Cherryhinton). The differences then, which separated the (probably) Belgic Iceni from the contemporary Girvii were presumably those of stature, and of external characters, such as the complexion, and the colour of the hair and eyes, the former tribes being ruddy in all probability, while the latter are supposed to have resembled the small dark-haired, dark-eyed Welsh and Shetland types. In conclusion it may not be out of place to append to this section of our account a sketch-map (no. I.) with indications of the localities whence the skulls in the University Museum were obtained, with a concise table shewing the general form of the skull in the several localities : and though exact dating is well-nigh impossible, it is fairly certain that the skulls referred to belong with but one or two exceptions to periods anterior to the Norman Conquest. With this our task is at an end. The hope is expressed, in conclusion, that this sketch either may stimulate some to examine the actual sites in the neighbourhood, or at least may provide definite reasons for visits to our University Museums of local Ethnology or of Anatomy.

LIST OF SOME OF THE PRINCIPAL SKULLS IN THE UNIVERSITY MUSEUM OF ANATOMY, OBTAINED FROM ANCIENT SITES IN THE COUNTY OF CAMBRIDGE, &c.

	Character of skull			Total
	1 Long, narrow	2 Medium	3 Short, broad	Total
Barrington	4	0	0	4
Bartlow	1	0	0	1
Burwell	1	2	0	3
Barnwell	2	1	0	3
Barton Road	1	0	0	1
Chesterton (gravel beds)	1	1	1	3
Huntingdon Rd (Roman site)	2	0	0	2
Jesus Lane	2	0	0	2
Various (Queens' Coll., S. John's Coll., Sidney St)	7	3	0	10
Castle Hill	1	1	0	2
Chippenham	1	0	0	1
Comberton (near Roman villa)	1	0	0	1
Fulbourn (War-Ditches)	3	2	0	5
Girton (ancient cemetery)	6	1	0	7
Haslingfield	1	0	0	1
Hauxton (coprolite digs.)	31	15	0	46
Kingston ,, ,,	11	6	2	19
Litlington	2	0	0	2
Madingley	1	0	0	1
Great Chesterford, Essex	1	0	1	2
Saffron Walden, Essex .	1	0	0	1
Brandon, Suffolk	23	23	5	51
Grand total...	103	55	9	167
Percentages...	61·6	33	5·4	100

Note:—The figures for the Brandon crania are admitted because this town is so near the border of Cambridgeshire as to be practically within the scope of this sketch. The figures are from measurements by Dr C. S. Myers of Caius College. They are thus to be clearly distinguished from the earlier figures which are estimates only, not supported by actual measurements, but which nevertheless will, it is believed, prove not far removed from the actual figures. The number of skulls of mean proportion will probably be increased when these estimates are checked by measurements, and the proportions will thus be nearer those obtained by Dr Myers for the Brandon skulls.

I.

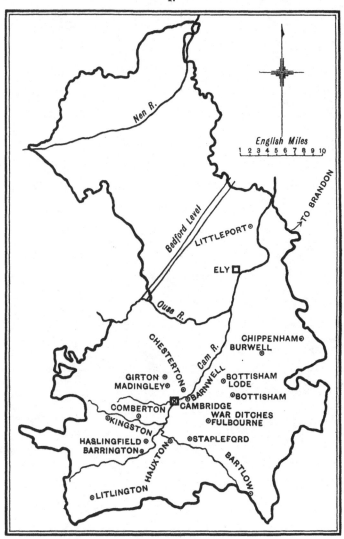

Map shewing localities in the County of Cambridge, whence ancient Skulls,
of Romano-British and East Anglian origin, have been obtained. The
specimens are now in the Anatomical Museum.

II.

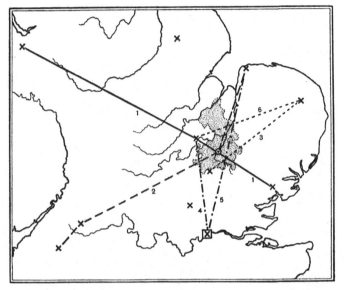

Six of the more important ancient roads crossing the County of
Cambridge:

1. Via Devana.   2. Akerman Street.
3. Icknield Way.   4. Ermine Street.
5. Pedder Way.   6. The Fen Road.

APPENDIX TO THE ARTICLE ON
THE MOLLUSCA.

Since going to press the writer is indebted to Mrs Hughes for the following additional localities and notes on living forms from her own observations.

Neritina fluviatilis; Hauxton Mill, Ditton.

Vivipara vivipara; the record on p. 123 is based on a single specimen in Wisbech Museum, which is possibly a young *V. contecta.*

Limnaea truncatula; Wisbech Canal (brackish water), River Nene, Roslyn Pit at Ely.

Planorbis crista; White House Drove near Littleport, March.

Vitrea rogersi; fairly common, Cambridge, Harston, Hauxton.

V. radiatula; Hauxton Mill.

Zonitoides excavatus; Madingley, March.

Euconulus fulvus; Grantchester, Madingley, etc.

Sphyradium edentulum; Roman Road on the Gogmagog.

Hygromia granulata; Hauxton, Whittlesford, Waterbeach.

Acanthinula aculeata; found again recently in Madingley Wood by Mrs Hughes and Mr Laidlaw (see p. 121).

Vallonia pulchella; Thorney, Cherryhinton, but less common than *V. costata,* which is found all over the county (see p. 122).

Helicigona lapicida; alive at Fen Ditton (see p. 130); seems dying out.

Helix hortensis; not common, but occurs at Fen Ditton, Haslingfield, Fulbourn, Cherryhinton and March.

Ena obscura; fairly common in and about Cambridge.

Caecilioides acicula; in holes in gravel, Cambridge Station, the Observatory, and alive at Hauxton (see p. 121).

Jaminia cylindracea; Bracondale (Cambridge), Babraham, Hauxton, Baitsbite, Duxford.

Vertigo pygmaea; Roman Road on the Gogmagog.

Balea perversa; on willows along Bourne Brook and on elms N.W. of March.

Clausilia laminata; Madingley, Haslingfield, Cherryhinton, Fen Ditton.

C. bidentata; Bracondale and Ravensworth (Cambridge), Trumpington, Cherryhinton, Toft, Whittlesford.

Dreissensia polymorpha; March, Roslyn Pit at Ely.

Unio pictorum; fairly common in Ouse and Cam below Cambridge.

Pisidium henslowianum; Hobson's Conduit, Coe Fen, Hauxton, Whittlesford.

Cardium edule (alive) and *Macoma balthica* (dead) in brackish waters of the Nene at Tydd Gotes.

INDEX.

CAMBRIDGE: PRINTED BY J. AND C. F. CLAY, AT THE UNIVERSITY PRESS.

Printed in the United States
By Bookmasters